Routledge Revivals

Transport Policy Problems at National and International Level

Transport Policy Problems at National and International Level

H. Osborne Mance

First published 1959 by Frank Cass and Company Limited

This edition first published in 2018 by Routledge
2 Park Square, Milton Park, Abingdon, Oxon, OX14 4RN
and by Routledge
52 Vanderbilt Avenue, New York, NY 10017, USA

Routledge is an imprint of the Taylor & Francis Group, an informa business

Publisher's Note

The publisher has gone to great lengths to ensure the quality of this reprint but points out that some imperfections in the original copies may be apparent.

Disclaimer

The publisher has made every effort to trace copyright holders and welcomes correspondence from those they have been unable to contact.

A Library of Congress record exists under ISBN:

ISBN 13: 978-0-367-13972-8 (hbk)
ISBN 13: 978-0-367-14404-3 (pbk)
ISBN 13: 978-0-429-03187-8 (ebk)

ITF

Transport Policy Problems
at
National and International Level

A contribution by the Transport Workers' Unions

with a

FOREWORD

by

Brigadier-General Sir H. Osborne Mance, K.B.E., C.B., C.M.G., D.S.O.

INTERNATIONAL TRANSPORT WORKERS' FEDERATION

Distributed by
FRANK CASS & CO. LTD., LONDON

First published 1959.

Published by

INTERNATIONAL TRANSPORT WORKERS' FEDERATION
Maritime House : OLD TOWN · CLAPHAM · LONDON S.W.4

Distributed by

FRANK CASS AND CO. LTD.
91 Southampton Row,
London, W.C.1

Printed in the United Kingdom by

GEO. MARSHALL & CO. LTD.
London, S.E.1

FOREWORD

Brigadier-General Sir H. Osborne Mance,
K.B.E., C.B., C.M.G., D.S.O., M.Inst.T.
President of the British Institute of Transport (1949-50) London
Former member of the Rhine Central Commission
Former member of the Transport and Communications Commission of the United Nations, New York

The impact of motor transport on the national transport systems was beginning to be felt at least thirty years ago. The work of numerous committees in the 1930s, while it led to the technical regulation of road transport and the control of road services by licensing, did not contribute any real objective study of the problem of coordination. This was due largely to the reluctance of vested interests to face up squarely to the issue of what their future roles should be, while Governments contented themselves with keeping the ring and administering first-aid when one or other party showed signs of bleeding to death.

All progress in coordination was suspended during the war with its critical shortage of transport. It is only in the last ten years that the problem of transport coordination has thrust itself into the foreground. By this time attempts at national coordination had been governed more by political than by technical considerations.

An important change in the situation was brought about through the need for international action arising out of political developments in Europe. As a result numerous international bodies have been set up to deal with transport. The members of these bodies are not representatives of particular vested interests. The growing evidence of the similarity of the transport problems in all countries has led to a far more objective study of the underlying principles of coordination than has been possible nationally. As a result of discussion great progress is being made towards the establishment of agreed principles and in the study of their implementation in different political and economic circumstances. It is, therefore, a good thing that the International Transport Workers' Federation, representing the workers in all forms of transport, should themselves undertake an enquiry into transport policy problems, and I should like to congratulate them on their informative and unusually objective study, with a great deal of which I am in complete agreement.

It is most important that the different elements of public opinion — transport undertakers, employees, users, and the taxpayer—which may be affected by the more detailed implementation of transport coordination, should be kept advised of the reasons for such developments as may occur. In performing this task as regards transport workers, the Federation has already taken a big step by the production of their study, which exhibits a remarkable degree of compromise between representatives of workers in the various forms of transport operating in countries with different economic outlooks on transport.

It is now well known that the problem of transport is approached on the one hand by those States and individuals who consider that in principle the coordination of transport should be based on purely commercial considerations, and on the other hand by those who consider transport to be a public service to be used for the fulfilment of national economic and even political policy. The study endeavours to hold the balance between these extremes and prefers indirect means of influence rather than direct controlling intervention as a means of implementing future policy.

The study accepts certain fundamental principles as the necessary basis for transport policy. While rejecting competition based solely on free enterprise it is proposed that competition should be governed by the price-cost mechanism under equal conditions within the framework of planned competition. Non-paying services imposed by the State for economic or political reasons should be compensated by the State. In this respect I would suggest that future studies may disclose the difference between non-paying services arising out of the specific preferences to certain classes of traffic, which should be subsidized by the State, and non-paying essential public services necessary to carry out the national policy for the distribution of population and industry.

The study comes down definitely in favour of free choice of the means of transport by the users, though it advocates the regulation of transport for own account. It also recommends a licensing system for each branch of the transport industry to prevent over-expansion of transport and to safeguard the maximum use of available transport capacity. The social conditions of the personnel should correspond with those of the leading industrial undertakings.

It is recommended that each form of transport should be self-supporting, including its track. On this point I have suggested elsewhere that the cheapest transport for the community as a whole would be furthered by the pooling of track costs, which would also have other advantages. The study appears to leave the way open for future consideration.

A further important recommendation is that the rates of all branches of transport should be officially approved and publicised, fixed rates being replaced by maximum and minimum rates. It is evident that with equal conditions of competition the alternatives are either no publication, or full publication by all forms of transport. The study fully recognises the difficulties of application of this latter course in the case of the small man and makes appropriate proposals for the rigid organization of non-scheduled road transport.

There are interesting paragraphs on new investment and disinvestment, with the recommendation that a central body should be entrusted with the task of balancing competing investments and establishing guiding principles for long-term plans of investment for the entire transport industry.

This, and other functions, would be carried out by a Transport Advisory Council with regional committees and special supervisory bodies. Any such organization would presumably have to take account of existing national machinery on the one hand, and also the possibility at some future date of the creation of an International Transport Authority so strongly recommended by the International Transport Workers' Federation since as far back as 1944.

The third part of the book is devoted to an examination of the transport problems arising out of the European Common Market. An important section deals with the effect on the wages and conditions of labour in a common transport market.

The concluding section reviews in some detail the developments necessary for a closer integration of European transport policy in a Common Market and a Free Trade Area, pointing out that coordination of the different systems of coordination will be difficult if different countries maintain their full sovereignty. Transport policy is at present in a state of rapid evolution and it will take some time to reconcile divergent national policies which, however, I feel may well turn out to be differences in the application of common underlying principles.

The statement that the study does not claim to be complete and could not possibly be complete shows a vivid appreciation of the present state of flux of transport thinking, which has yet to tackle many important problems, to the solution of which the International Transport Workers' Federation can make a valuable contribution.

11TH NOVEMBER 1958.

TABLE OF CONTENTS

INTRODUCTION

In presenting this study of transport problems to its affiliated unions and thereby to larger circles of the public, the I.T.F. feels entitled to point out that it has been paying particular attention to these problems for a number of years. The question of coordination of the branches of the transport industry was discussed as long ago as the end of the 'twenties. Proposals concerning the division of functions of the transport industry were for the first time adopted on the occasion of an International Railwaymen's Conference at Madrid in the spring of 1930. The first efforts aiming at an integration of European transport are of an even earlier date. The 1924 Congress of the I.T.F. at Hamburg adopted a resolution on the creation of the United States of Europe. The then General Secretary of the I.T.F., Edo Fimmen, subsequently published a Manifesto entitled "United States of Europe or Europe Co. Ltd." which was devoted to the international task of the trade unions in this field.

Numerous reports and resolutions concerning national and international transport problems have been discussed and approved since these early endeavours. Even the Second World War could not stop these efforts of the I.T.F.; on the contrary, the first basic principles of the structure of a post-war European transport industry were formulated 1943/44 as a result of the comprehensive discussions within a special transport committee. They were approved by the Executive Committee of the I.T.F. in 1944 and by Congress at Zürich in 1946.

Our survey of the efforts of the I.T.F. throughout the years would be incomplete without a mention of the useful cooperation with the International Labour Organization in this field. Among its many important achievements since the Second World War we only wish to refer in this connection to the adoption of the resolution on labour problems arising out of the coordination of transport adopted by the 1951 Inland Transport Conference of the I.L.O. at Nervi. The importance of the principles embodied in that document is enhanced particularly by the fact that it was adopted almost unanimously by the delegates of governments, employers and workers. The text of that Resolution is reproduced in an Annex to the present study.

In accordance with a decision of a Conference of the inland transport sections of the I.T.F. in autumn 1955 a committee of experts was entrusted with the task of studying the problems of coordination of transport at national and international level under the auspices of the secretariat of the I.T.F. and to try to make available to affiliated unions certain guiding lines for a policy to be pursued in this field. The 1956 Congress of the I.T.F. at Vienna confirmed these decisions and also adopted a comprehensive resolution on the creation of a common European transport market.

The present study represents a further result of our efforts and of the instructions given to the committee of experts in 1955 and 1956. That committee consisted of one representative each of the French Force-Ouvrière Union of Public Servants and Transport Workers (M. Gilbert, lic-ès-sc.), the Netherlands Transport Workers' Union (Drs. P. Seton) and the German Railwaymen's Union (W. Mikkelsen, B.A. Econ.) as well as the German Public Servants' and Transport Workers' Union (Dr. K. Kühne). The meetings of the committee always took place under the chairmanship of Mr. H. Imhof, Section Secretary. The translations of the German original into English, French and Swedish were carried out by the secretariat of the I.T.F., assisted by Dr. Kühne, Stuttgart (French and English) and Dr. Meidner, Stockholm (Swedish). The draft reports were the subject of preliminary discussion by the competent bodies of the I.T.F. The Amsterdam Congress of the I.T.F. from 23-31 July 1958, unanimously adopted the report as a basis for the future efforts of the affiliated unions in the field of transport policy.

Anybody who has ever been concerned in any way with problems of the transport industry will realize the extent of the difficulties which opposed our joint endeavours. They arose wherever an attempt was made to create some order and to secure for each branch of the transport industry the share of traffic to which it is entitled in view of the particular advantages which it can offer; they also arose whenever a sufficiently broad basis of competition had to be found. These difficulties were certainly by no means reduced by the fact that the committee consisted of experts belonging to unions in different countries representing the interests of the personnel of various branches of the transport industry. The texts are consequently in many instances based on compromises which frequently could only be arrived at after several unsuccessful attempts.

It is therefore only fair to request that these difficulties and the available time between autumn 1955 and spring 1958—a rather limited period in view of the innumerable problems which had to be discussed—be duly taken into consideration when judging the merits of this study. It makes no claim to perfection and is not intended to imply any obligation on our affiliated unions to adhere to rigid guiding lines.

The most important part of the study is Part One, which deals with the problems of coordination of inland transport. Particular attention was devoted to that part by the Transport Advisory Committee and subsequently by the competent bodies of the I.T.F. Their objective was to find a conception which would combine a healthy competition with the necessary direction and supervision, based on the realities of the political and economic conditions within Europe.

The special problems which are treated in the second part of the study have, in the main, been merely indicated. The more comprehensive considerations of these problems and the determination of agreed expressions of opinion will have to be left to our affiliated unions. Neither a thorough examination of the problems of urban transport nor of those of civil aviation or the port industry would be feasible without at the same time investigating the various basic problems which are involved.

The third part of the study is connected with our efforts aiming at the creation of a common European transport market, as expressed most recently in the above-mentioned Resolution of the Vienna Congress of the I.T.F. 1956. We have, however, refrained from setting up the first milestones on the road to this still very distant objective and have tried instead to enumerate the problems arising from the reality of the Treaty on the Economic Community of the six countries concerned.

As already mentioned, our study does not claim to be complete, and could not possibly be complete. Time obviously does not stand still. It will aways be necessary for us as trade unionists together with all others who stand for conditions of order and security within an efficient transport industry to adapt ourselves to developments and to try to find appropriate solutions. To this end, our work is intended to be the foundation on which we can continue to build. It should serve as a guiding line to our affiliated unions in questions of transport policy and at the same time as a framework of a basic trade union conception. It is also intended to be a contribution of the free trade unions to the discussions of today's problems of the transport industry.

In conclusion we should like to express our sincere thanks to all who have assisted us in the preparation of this document, in the very first instance to our affiliated unions which have enabled their experts to devote part of their time to our work, then to the experts themselves who had accepted this task and finally to the competent bodies of the I.T.F. whose difficult task it was to examine the draft reports in preliminary discussions. The purpose of these joint efforts was to try to reach the common objective of the affiliated unions of the I.T.F., namely, to assist in the creation of sound conditions within the transport industry and thereby to serve the interests of the nations, the economies and all who work in transport.

OMER BECU,
General Secretary

London, August 1958

PART ONE

Problems of the Coordination of Inland Transport

PREAMBLE

A. Historical Development and Transport Policy

Transport has always played an extremely significant part in all economic activities. The development of transport laid the foundation for the industrial revolution, which gave birth to gigantic concentrations of industrial plants in the coal regions, attracting vast labour forces. These powerful industrial concentrations demanded an improvement in transport, as on the one hand the feeding of these large numbers of people had to be secured, and on the other hand the raw materials destined for distant manufacturing centres had to be taken there. The transport of masses of workers from their homes to their place of work and back imposed hitherto unknown tasks on the passenger services. In Europe all this led to an increasing degree of division of work, both within the individual industries and between the different regions. The economy of to-day would be inconceivable without the existence of the modern means of transport.

As little as half a century ago travel was considered to be more or less of a luxury, which only privileged classes could afford. To-day, however, transport has become a necessity for more classes of users, with the same claim for satisfaction as the demand for electricity or postal services.

The transport industry is therefore a branch of economy of equal importance as agriculture, industry or commerce. This fact has not always been properly taken into account in the economic policies of the states. Transport has often been regarded only as a means for attaining certain economic and social-political objectives. The choice of sites of the production centres of the national economy and the prices of their products were consciously influenced via the tariff structures of the different branches of transport, above all the railways. Investments in transport have also been planned from the point of view of a policy of full employment.

In the development stage of transport tracks were constructed and maintained merely to satisfy natural transport requirements. Transport policy was only required to adapt itself passively to the existing geographical conditions. Only in modern times, for reasons of strategy and economic and social policy, the opening up of new regions was undertaken by laying new tracks. In this way transport became more and more an accessory of the economic and location policy, developing from its hitherto passive acceptance of locational conditions into an active element assisting in the attainment of locational objectives.

There can be little doubt that in modern times, with the far-reaching commercial penetration of provincial areas and consequent extensive scattering of industrial enterprises, the possibilities and necessities of formulating and influencing a policy for the attainment of locational objectives are reduced. In addition, the production of ubiquitous raw materials and the decreasing share of raw materials in the production of many goods lines have increasingly lessened the need for supplying outlying areas with cheap raw materials. Assisted by the advent of new means of transport, a completely new position resulted for the railways. Previously it was thought that the railways, as flourishing enterprises, could subsidize distressed areas. Now the changed competitive position deprived the railways of their monopolies, whereas at the same time, the obligations imposed on the railways remained unchanged. They therefore became themselves an ailing branch of the economy, no longer capable of subsidizing other distressed industries.

The rapid progress made in the field of motor-transport brought about such a development of road transport that a constantly increasing proportion of the total traffic was lost by the railways. The railways' deficit then grew from year to year. The increase in this deficit, which has to be covered by the community, has caused the general public to pay increased attention to the problem of the transport industry. This is the cause of constant interventions in this field, with the object of reducing the deficit of the railways. It is, however, true that in spite of all efforts made so far, this deficit has only disappeared in very few instances, and that the problem of coordination of transport at any rate exists as much as ever.

In view of this situation the question arises whether the present subordination of transport policy to principles of locational and general economic policy may be maintained. There are many indications that the time has come to operate transport undertakings more in accordance with commercial principles and the conceptions of economic self-sufficiency. Modern economic policy has such a multitude of means at its disposal, that measures in the field of transport—and particularly tariff policy—for the attainment of objectives of the national economy in general may largely be dispensed with. Nevertheless, the industry should also in future be considered a public service. Even with the greater stress on commercial principles the fundamental principle of ensuring that all engaged in any field of human activity are provided with an efficient transport service may not be lost sight of.

B. Public Character of Transport

The possibilities of actively intervening in the sphere of transport are explained and justified by the fact that the transport industry is largely a public service. This public character is based on several considerations.

The determination and satisfaction of collective requirements or those of the entire community are tasks incumbent on the State and the public authorities. Both are intrinsic tasks of the transport industry. Furthermore, the development of transport forms the basis of the development of the entire national economy. Consequently, disregarding varying conditions of ownership, for these two reasons alone, the transport industry as an economic factor of the first order should be subject to the influence of the State.

Wherever a unification of the entire transport industry has already been obtained (e.g. centralization of control, nationalization, and similar methods) economic planning measures are simplified to a certain extent. This position does, however, not exist in the majority of countries. Thus, the dualism of public and private ownership in transport creates basic and organic problems within the framework of the national economic policy.

The trends of economic policy supply one of the most important reasons for the determination of the public character of the transport industry. Due to its obligations towards the community the possibilities of the transport industry to exploit favourable economic situations are in many instances more limited than those of other industries. The stage may even be reached where in times of an economic boom, the remaining industries are subsidized by the transport industry. On the other hand, the transport industry is particularly vulnerable to economic depressions which very frequently make themselves felt there in the first place. The business cycle policy of the State must consequently include planning measures, in order to counterbalance the vulnerability of transport to cyclical fluctuations. At the same time investments in the transport industry lend themselves particularly well to the implementation of a policy of full employment.

A further special task of the State consists in safeguarding the interests of the community, even beyond those of transport users. From a usufructuary of transport, man may become its victim both as a transport user and a worker and finally as an outsider. The problem of safety admittedly belongs to the general police duties of the State; it does, however, exert a decisive influence on economic and social questions. Safety measures for the protection of both those employed in transport and for the travelling public as a whole may, to a large extent, solely be guaranteed by statutory regulation.

By its very nature, transport shows a tendency towards traffic monopolies and differentiated services. This tendency led already in the early days to the realization of the need for the supervision of freight rates by the State as representative of the interests of the community. By further application of this principle, we arrive at the conclusion that all means of transport must be submitted to such control to an equal extent in order to prevent the community being exploited by relatively powerful pressure groups in the transport industry. Political decisions which affect the transport industry are—although in a varying degree—of a certain importance for the choice of the sites of industries. Site-bound industries require the creation and improvement of communications if the harmonic development of the entire economy is to be safeguarded. This is of particular importance with a view to encouraging the division of labour and creating a healthy competition.

The obligations to which the transport industry is submitted regarding the fixation of transport rates in order to safeguard the public interest are of particular importance to the national financial policy. They may, on the one hand, cause an increase in the real income of the community; on the other hand, there is the danger of these obligations exerting an undesirable influence on income conditions in the transport industry itself. These developments may be caused by cartel-like rate agreements whereby certain undertakings would reap larger profits than would be the case under conditions of free competition. From a long-term point of view, similar conditions may on the other hand attract enterprises to certain transport regions and thereby cause over-equipment and over-development. The realization of this danger leads as a conclusion to the necessity of extending the afore-mentioned obligations beyond the sector of prices.

The need for influencing incomes impinges on wage policy. It may and must, however, on no account lead to a reversal of the principle of collective wage negotiations. Similar considerations apply to the regulation of working conditions. Wherever any special considerations from the point of view of social policy exert any influence on the transport industry they should not be allowed to distort the competitive position. It is, however, quite conceivable that a neglect of the social aspects, leading to unfair competition at the expense of the workers may cause such a distortion.

Similar considerations apply particularly to the policy connected with the opening-up of potential industrial areas. This should be considered as a task devolving on the State and should not be allowed to become a charge on the different branches of the transport industry which could not be justified on grounds of business economics.

The above considerations show the need for the State to influence transport beyond the normal influence which it exerts on the economy. This influence should, in the interest of preserving the greatest possible flexibility and adaptability of the transport industry, be concentrated on the establishment of fundamental principles. A too detailed scheme will be open to the criticism of over-planning and may be detrimental to the remunerativeness and productivity of the transport undertakings.

C. New Ways of Inducing Maximum Productivity

It is in fact clear that in the past public influence in many instances began to extend beyond its actual purpose. At a time when the transport industry has outgrown the monopolistic rigidity of the earlier railway policy, it appears impossible to ignore the competition between the forms of transport and the concomitant problems of the structure of market conditions and costs.

This does not mean that an efficient co-ordination of transport and the future avoidance of uneconomic investments could not also be achieved by nationalizing this industry and/or by imposing a unified administration on all transport undertakings. If the I.T.F. does not recommend this solution in this memorandum, it is above all because this would encroach on political aspects. It is, therefore, a question of finding a middle road between the extremes of a planned transport industry and its complete liberalization. The problem of coordination of transport will, in any case, continue to exist even in a nationalized transport industry.

This attempt has been made in the chapters which follow. Wherever the choice presented itself between direct controlling intervention and indirect means of influence, the latter has largely been given preference. This method was born of the conviction that the preservation of the greatest possible measure of flexibility can in general only be achieved if the cost structure is chosen as the foundation of a flexible price policy, taking into due consideration the actual market conditions. Hitherto the sharp contrast between a rigid price structure in transport and the fluctuating price development in most other industries constantly conjured up the threat of discrepancies and distortions. The possibilities of increases in the productivity of the transport industry were limited by the artificial restriction of investments. This restriction was largely expressed by an inability to plan investments a longer time ahead as long as it was not known to which extent transport undertakings would be freed of extraneous and political charges. The development of added value inevitably lagged behind that of industry as a whole. This could not help producing repercussions on the development of wages and working conditions. While the physical and psychological demands of modern transport on those whom it employs have constantly been increased, stagnating wages have made it increasingly difficult to attract new blood and to safeguard a high level of productivity.

This lagging behind of the transport industry in comparison with other industries has consequently caused a danger of a similar lagging behind in quality, whereas, for the very reason of these increased requirements, selection of the most able personnel should be the most urgent task of the transport industry.

It would appear that these difficulties can only be overcome if the present chaotic position which opposes the individual branches of transport to each other is eliminated and a way is found to a maximum coordination of the different branches of the transport industry.

The introduction of new transport methods in goods transport will incidentally initiate a closer cooperation. Containers and pallets for general goods- and bulk-traffic and combined road-rail transport (piggyback) for full loads may simplify operations and at the same time make them cheaper and faster. These advantages can only be fully utilized if the new transport technique is adopted by all branches of the transport industry and if a common pool is set up to administer the technical resources. Provision should further be made for the sponsoring and further development of the transport technique in the interest of the national economy by means of an appropriate tariff and investment policy.

The present report endeavours first of all to determine the measures which in our opinion are required for a coordination of transport. In the preceding chapter, we have briefly summarized the reasons why transport as a whole has to be considered as a public service. The same reasons apply to economic areas within national boundaries as well as to integrated areas.

As far as the solution of the various problems is concerned. the creation of a common European market may result in certain modifications primarily due to the fact that the assimilation between the participating countries must take place within the most diverse fields of competence of the economic, fiscal and social policy. We are, however, of the opinion that mutually agreed measures for coordinating transport within the various European countries would facilitate the creation of an overall market for a free exchange of goods and services. With regard to a common market endeavours should be made to arrive, if possible, at uniform European regulations which could be applied to all branches of the transport industry. Competition in transport beyond intervening frontiers without any distortion of competitive conditions is only possible if an adaptation of the conditions on which such competition is based can be obtained beforehand within the transport industries of the different countries.

I. BASIC PRINCIPLES OF TRANSPORT COORDINATION

The I.T.F. maintains that the coordination of the means of inland transport should aim at a maximum satisfaction of the transport requirements of the national economy. This means in the first instance regular and reliable transport services of high quality combined with the lowest possible expenditure for the national economy. Any development should be avoided which results in the transport industry and its branches being used as instruments for the fulfilment of tasks which are outside the framework of industrial self-sufficiency. There are other means at the disposal of the State for the purpose of sponsoring economic activity, determination of the location of industries and implementation of social tasks which must first be exhausted. We do, however, recognize the fairly widespread practice to facilitate the satisfaction of transport requirements of commuters, students and other circles of the population because of general considerations of social policy by means of special facilities. It is particularly with reference to these requirements that we believe that the supervision and direction exercised by the state should be more extensive in scheduled passenger transport than in goods transport. This gradual differentiation of our coordinating proposals is intended to indicate that they are governed exclusively by practical considerations which not only take into account the overall economic interests of the community but also the objectives of modern social policy, the raising of the general standard of living and in particular that of transport workers. We realize that there is no universal panacea for a maximum satisfaction of all transport requirements of the national economy and that this objective must be continuously adapted to the circumstances prevailing at any given time.

Any system of coordination which precludes arbitrary measures should imply the free choice of the means of transport by the user. An influence on such choice may only be exerted by means of the price-cost mechanism. Any purely officially-controlled division of the tasks of the different branches of the transport industry which is merely based on cost considerations without taking into account conditions on the market involves the risk of arbitrary decisions and may easily lead to an ossification of the entire transport industry.

It is, on the other hand, realized within the I.T.F. that the best possible satisfaction of the transport requirements of the national economy cannot be guaranteed by a competition between the means of transport based on the principles of free enterprise. Any form of economy based on free enterprise presupposes conditions of perfect competition which, however, in the field of transport, exist neither so far as the supply nor the demand is concerned. Whereas the railways, in view of their extensive capital requirements, show a tendency towards large-scale undertakings and monopolies, road haulage, with the relatively small outlay of capital required for this trade, shows a marked tendency towards an atomistic supply of transport services. This fragmentation not only encourages a periodically recurring tendency towards a ruinous competition but also largely prevents any coverage by progressive agreements which are the objective of transport workers' unions. The mutual effects are equally negative both for the transport industry and those employed in it. Conversely, the offer of transport services in inland navigation is characterized by a parallelism of atomistic competition exercised by innumerable individual undertakings and oligopolistic competition between big shipping companies.

Furthermore, the services which form the subject of supply and demand on the transport markets are no homogeneous products but are only interchangeable within certain limits. This is due to the fact that in the case of particular types of goods special factors (speed, door-to-door services, sensitiveness of the goods, etc.) are in certain circumstances more important than financial considerations, which explains why certain customers prefer a particular means of transport.

In view of this fact, each branch of the transport industry operates on a very great number of partial markets where, apart from an imperfect competition between an almost unlimited number of undertakings, all characteristics of monopolistic and semi-monopolistic market structures may be found. In these partial markets transport services are in many instances not only offered by undertakings of one and the same branch of the transport industry but there is also competition between undertakings of different branches.

The offers of transport services also lack the elasticity required by an economy based on free enterprise by virtue of the particular conditions of production which apply. The balance between supply and demand for transport services meets difficulties particularly due to the fact that transport undertakings, unlike those of manufacturing industries, are not in a position to stock-pile. It is due to this fact that, above all, transport undertakings which are subject to the obligation to act as common carrier and to maintain services are compelled to adapt their operating capacity to the demands of seasonal peak requirements. The more each branch of the transport industry is compelled by its technical characteristics to invest in fixed installations the greater these difficulties will become. In the majority of cases, these installations may only serve one single purpose and consequently the possibilities of switching over to the production of other goods or services in times of a declining demand for transport services are very restricted indeed.

Similarly, so far as the demand for transport services is concerned, no uniform structure of competition is apparent. It is as a rule both split up into innumerable small components and concentrated in the hands of a limited number of agents of the trading public (forwarding agents) who in many instances occupy a position of strength in their negotiations with transport undertakings. Since these large firms in the various countries frequently act as carriers themselves, certain branches of the transport industry may thereby receive preferential treatment. In such cases the result will be a limitation of supply and demand or rather a link-up between demand and certain forms of supply. Such ties may also exist in other forms, for instance in the form of private sidings. The demand will, on the other hand, be in a comparatively weaker position mainly in remoter areas where the demand for transport services is split up in innumerable fragments, e.g., in the case of mixed goods and has to contend with a number of carriers.

On the whole it may be said that there is a lack of elasticity of the supply of transport services which corresponds to a lack of elasticity of the demand for such services. Changes of rates of certain transport undertakings which do not exert any far-reaching influence on the tariff structure may admittedly cause quantitative changes of the volume of traffic available to the individual undertakings but will not influence the overall demand of the national economy for transport services from a purely quantitative point of view to any noticeable extent. This overall demand depends on the quantitative overall demand for goods of industrial production which is affected by other factors to a very much larger extent than by the cost of transport. Rates can, however, cause noticeable changes in transport distances.

In case of a decline in the demand for transport services, all these circumstances will cause an abnormal reaction of the transport undertakings so far as their quantitative supply of transport capacity are concerned. Experience has shown that a recession in the demand produces a tendency towards maintaining a constant supply of transport capacity, which must finally exert a pressure in the direction of rather debatable rate reductions. Considered from a long-term point of view, freight rates in general must, as a result of mutual competition, sink below the level of the total average costs.

The tendency towards a ruinous competition in transport is still further intensified by the differences in the quantities of goods which are carried between two given points at any time and by the encroachment on commercial transport by transport for own account. These factors exert a negative influence on the social conditions and are largely responsible for the fact that social progress in certain branches of the transport industry has lagged behind the general development.

Because of these fundamental considerations we are convinced that the price-cost mechanism as an instrument of coordination of European goods transport may only exert a regulating influence within the framework of a planned competition. Only a competition which is subject to flexible regulation but at the same time strictly supervised by the State is economically useful and in the public interest. It also offers the only possibility to eliminate the factors which distort the prerequisites of a fair competition.

The European transport workers' unions believe that a reorganization of the transport industry is only feasible in the present circumstances if measures adopted by the State for the purpose of regulating and supervising competition duly take into consideration the following

PRINCIPLES:

1. Competition should be based on the natural differences between technical and economic efficiency of transport undertakings.

2. Each branch of the transport industry should in principle aim at economic self-sufficiency.

   Arrangements should be made for the public transport authorities to maintain balance sheets and profit and loss accounts for each branch of the transport industry by means of appropriate accounting methods which contain all elements required for determining the degree of economic self-sufficiency of each branch of the transport industry.

   Branches of the transport industry or individual carriers which have to apply rates and/or carry out tasks imposed by the State for economic or political reasons and which do not cover appropriate costs shall be compensated accordingly. Financial compensation granted by the State should be accounted for as receipts of the branch of the transport industry concerned.

3. Social conditions of the personnel of the transport industry should not be allowed to lag behind the general social development. Wherever modern transport involves special efforts by the personnel concerned social conditions should be above the average level of those in the leading industrial undertakings.

4. In order to prevent an overexpansion of transport and in order to safeguard the maximum use of the available transport capacity, a system of concessions and/or permits shall be introduced for each branch of the transport industry.

5. The competent bodies shall apply a rates policy which takes into account the economic and technical advantages of each means of transport and thereby induces transport users *via* the price mechanism to avail themselves of the branches of the transport industry which are most useful in the overall interests of the national economy.

6. Investments in tracks, installations and vehicles should be coordinated in a manner which takes into account present and future transport requirements. Investments with the objective of increasing safety of operation and developing new methods of operation and transport should receive priority.

7. Regulating and supervisory measures adopted in the field of transport policy shall embrace transport for own account in order to limit its restricting influence on public and commercial transport and to prevent it competing with means of collective transport by means of below cost-level rates.
8. The authorities responsible for transport policy shall be assisted by bodies the composition of which should reflect the interests of all responsible parties engaged in transport. The application of the legislation governing coordination of transport shall be safeguarded by appropriate supervisory measures.

The special conditions obtaining in scheduled passenger transport and particularly the requirements of commuters' and "social" traffic during peak hours justify controls within this branch of the transport industry which cause more far-reaching repercussions than those outlined above. In many instances the safe and reliable operation of these services will be incompatible with the principle of economic self-sufficiency even if reductions of fares are compensated by public grants; the prerequisites of a healthy competition consequently no longer exist. In similar cases it would be preferable to grant certain carriers priority for handling this type of traffic.

Individual transport which shows a continuous increase not only competes with scheduled and non-scheduled passenger transport but also presents an obstacle of ever-growing dimensions. In the case of road haulage, too, the congestion of the roads and particularly the bottlenecks within built-up areas constitute a considerable hindrance.

On the other hand, the fact that the private car is increasingly becoming a means of transport of large parts of the population constitutes a favourable development as a symptom of an increase of the general welfare. Not even the increasing lack of space within which transport can operate will be able to halt this development.

In order to reconcile the opposing interests and in the interest of the safety of operation, care should be taken to prevent an increase of the traffic chaos already prevailing within the built-up areas which may be brought about by the one-sided extension of the network of roads outside the towns. If improvements of the urban network of roads are not feasible because of architectural, financial and other reasons and the construction of super-imposed highways appears impracticable, the danger of a relative overloading of urban areas will be increased. The only efficient way to meet this situation consists in subjecting individual transport as well as goods transport within the towns to certain limitations so far as time and area of operation are concerned and by simultaneously granting priority to public transport for the purpose of handling mass transport. Above all, a coordination of regulating measures with regard to traffic problems and town planning measures will be necessary.

II. ELIMINATION OF THE DISTORTIONS OF COMPETITION AND EQUALIZATION OF OPERATING CONDITIONS

At the present time, the various branches of the transport industry are subject to widely differing and partly abnormal conditions of competition. This lack of balance which is caused thereby makes it impossible to place all branches of the transport industry on an equal footing within a regulated competition, an objective which is generally advocated within the transport industry by the unions affiliated to the I.T.F. If, however, conditions of ownership as they nowadays exist in the transport industries of most countries are intended to remain unchanged in principle, ways and means will have to be found in order to harmonize the far-reaching obligations of public transport undertakings with the relative freedom of private undertakings. Measures of this kind have to be adopted before the initiation of the actual coordination of the branches of the transport industry because they have to pave the way for such coordination in the first instance.

In connection with this problem we only wish to consider the following main aspects:—

- (A) social conditions;
- (B) obligations imposed by the State:
  - (1) the cost of the tracks to be maintained and to be constructed;
  - (2) taxation of transport undertakings;
  - (3) conditions governing the obligations to adhere to officially approved rates, to act as common carrier and to maintain operation;
  - (4) extraneous and political charges;
  - (5) conditions of civil liability;
- (C) the problems of subsidies:
  - (1) unremunerative rates;
  - (2) State refunds and grants.

A. Social Conditions

A distortion of the natural conditions of competition between the individual undertakings of the transport industry is primarily caused by the fact that the working conditions in comparable functions of the various branches of the transport industry still differ very widely. In order to arrive at a genuine equalization of the conditions of competition, it is indispensable to eliminate these differences so far as possible. A peculiar aspect of the equalization of working conditions in road transport and inland navigation is the frequent incidence of family undertakings. There is a danger of such family undertakings being in a position to carry out transport operations at rates which do not fully cover total costs by working excessively long hours, paying subnormal wages, and evading contributions to social insurance and thereby distort competition. These tendencies may be largely counteracted by the amalgamation of small undertakings in organizations, by fixing of minimum rates and by imposing compulsory social insurance.

B. Obligations Imposed by the State

1. Cost of tracks to be maintained and to be constructed

So far as track costs are concerned, the branches of the transport industry shall in principle cover from their own resources the current expenditure as well as the costs of production of new infrastructures required by them including interest on capital and their share of the cost of traffic police. This does not exclude the possibility of borrowings for the purpose of building tracks.

The practical application of this principle in road transport will encounter certain difficulties because the share of overall costs to be charged to individual transport must be taken into consideration. Such calculations of the share of costs have to take into account wear and tear of road surfaces and the occupation of road space. In our opinion, road transport undertakings and transport for own account should be charged with the share of expenditure which is caused by them. On the other hand, the special taxation imposed on these undertakings may be considered as their contribution towards the cost of roads and should be credited as such to the road accounts.

In inland navigation these difficulties are even greater. If the same principles were applied in order to arrive at canal charges which would completely reflect costs, inland navigation might be paralyzed on extensive stretches of the canal network. A solution of this kind is furthermore impeded by inter-governmental agreements, which provide for freedom from taxation and charges on international waterways. Irrigation systems, hydro-electric power stations and agricultural interests also have to be considered in this connection and prevent an exact assessment of the share of costs to be charged to inland navigation. It will therefore be extremely difficult to forcibly combine traditional concepts of transport with any economic scheme which correctly reflects the cost structure.

In view of these difficulties, any proposals which aim at relieving all branches of the transport industry of a given share of track costs by charging them to the community may only be considered as simplifications on broad lines. We believe, however, that our proposals, which provide for the application of the principle of self-sufficiency also with regard to track costs, are in the long run preferable to such generalizations even if considerable compromises have to be accepted in this connection.

2. Taxation of transport undertakings

Although a unification of the various systems of taxation is impeded by the problem of structural differences between the various branches of the transport industry, it should nevertheless be attempted. In principle, taxation must be unbiased and not be allowed to exert any influence on competition. It should, however, take into consideration the extent to which public obligations of an undertaking influence its budget compared with undertakings which are not subject to such obligations.

Contributions towards the cost of tracks in so far as they are embodied in the general scheme of taxation should be earmarked accordingly as definite budgetary allocations; always, however, with the proviso that in case of an economic recession more funds may be allocated than have actually been earmarked for that purpose; yet the commercial consequences of similar measures as well as those in the field of taxation should not result in unfair charges on the branches of the transport industry once the economic equilibrium is restored.

3. Obligation to adhere to officially approved rates, to carry and to maintain operation

The differences regarding the obligation to adhere to officially approved rates, to carry and to maintain operation exert an unfavourable influence on the conditions of competition.

In order to arrive at an equalization of conditions so far as rate obligations are concerned, all branches of the transport industry should be uniformly subject to the obligation to have their rates officially approved and to publicize them.

Regarding the obligation to carry, all scheduled transport undertakings should be subject to equal obligations of this kind.

Similarly, the obligation to maintain operation in so far as it is prescribed for scheduled transport should be applied in a uniform manner and regulated for each line from an overall point of view which would be valid for all branches of the transport industry.

An equalization of the conditions of competition in non-scheduled traffic (tramp traffic) may be facilitated by means of freight agencies. These agencies would not only arrange contacts between the trading public and the tramp carrier but also lend practical support to the application of equal rates and the unification of transport conditions. In addition the reduced amount of empty running which would be obtained thereby would contribute to an increase in productivity of this branch of the transport industry.

4. Extraneous and political charges

The chaotic conditions of competition, war damage and indirect repercussions of the war, as well as the extraneous charges which are in many cases imposed on public transport have caused considerable deficits on the majority of European railways and other public transport undertakings. Due to lack of sufficient reserves, these undertakings are, generally speaking, not in a position to finance from their own resources the reconstruction of installations destroyed during the war and the renovation and modernization of obsolescent plant.

In our opinion a remedy may be found by either imposing public charges equally to all branches of the transport industry or, wherever they are arbitrarily imposed on public transport undertakings by adequately compensating them by other means. This aspect deserves all the more consideration in view of the fact that a complete elimination of public obligations which have become established in the course of time is impossible.

We furthermore consider an adequate compensation for war damage and indirect repercussions of war as well as financial obligations resulting therefrom as indispensable and just. The covering of deficits of public transport undertakings by the State, which is practised in numerous countries, is intrinsically different from a compensation of extraneous charges. It does not take into consideration actual requirements and exerts an unfavourable influence on the sense of responsibility of managements and on advance planning of investments. A comprehensive financial reorganization primarily of European railways ought therefore to be carried out in the interests of a rational transport policy and a general elimination of out-dated practices.

5. Conditions of civil liability

In order to arrive at an assimilation of the conditions of competition, a unification of the regulations governing civil liability of all branches of the transport industry is indispensable except where natural differences justify deviations from this rule. Particularly drivers of vehicles should be exempted in a uniform manner from the consequences of civil liability, in order to prevent any risks arising from competition between rail and road being passed on to the workers.

C. The Problem of Subsidies

1. Unremunerative rates

This definition refers to rates which are imposed by the State on public transport undertakings in the interest of objectives which, although they may be justified from the point of view of the economic and social policy entail a loss of income for the transport undertakings concerned. The assistance granted to economically under-developed areas, certain industries or undertakings, by means of particularly reduced rates as well as extensive reductions of fares in commuters' and students' traffic is only compatible with competition if the undertaking concerned is compensated for the loss of income caused by the application of these reduced rates. Such losses of income are not always inherently equivalent to the difference between the receipts obtained by the application of reduced rates and normal rates. There are certain aspects, particularly of commuters' and similar traffic to be considered, such as good-will, rebates on large numbers and frequently acceptance of sub-normal standard of comfort which largely counter-balance the loss of receipts. The loss to be compensated should, however, cover at least the difference up to and including the entire prime costs.

In the course of the adaptation of public transport undertakings to the changed conditions of competition the question whether certain unremunerative rates could not be usefully replaced by planned direct subsidies ought to be examined.

The reason why this problem of subsidized tariffs exerts such great influence on the distortion of the conditions of competition is the fact that the types of transport which are in need of subsidies are of varying importance for each branch of the transport industry. If consequently the loss of income resulting from the application of these tariffs is fully covered by the State, there will be no distortion of the conditions of competition. Conversely, if there is only partial compensation and the undertaking concerned has to rely for the remainder on internal measures, the application of such rates will continue to cause a distortion of the conditions of competition.

2. State refunds and grants

Refunds granted by the State as compensation for extraneous charges or in certain instances also as compensation for special duties in the nature of a public service are compatible with the principles of economic self-sufficiency and equalization of the conditions of competition provided they are granted to all undertakings which are subject to similar obligations in accordance with identical principles.

The same consideration is applicable to subsidies of the State which are granted within the framework of an anti-cyclical policy, *i.e.* a policy which involves an intensification of investments in times of a threatening economic recession, because such subsidies also do not contribute to any distortion of the conditions of competition, provided they are granted to all branches of the transport industry to an equal extent. Neither will subsidies by the State which are granted to all transport undertakings in the interest of technical progress contribute to a distortion of the conditions of competition. They should, however, be undisguised and should be withdrawn as soon as the envisaged purpose is fulfilled.

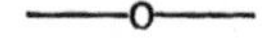

The unions affiliated to the I.T.F. believe that the application of the aforementioned principles may largely lead to an elimination of the initial differences between the conditions which govern competition between the branches of transport industry. They are of additional importance for scheduled passenger transport which is coordinated in accordance with more strictly applied regulating principles because they would eliminate a number of undesirable possibilities of subsidies which come into play in the case of unequal operating conditions and the disturbing influence of which is frequently underestimated.

If, for instance, certain scheduled transport undertakings do not cover any or not the entire costs of tracks used by them, these undertakings and/or the transport users which avail themselves of their services enjoy a differential at the expense of the community. Conversely the community will always reap an unearned benefit at the expense of the transport undertakings and their personnel and/or the transport users if these undertakings are charged with special taxes beyond their own particular contributions.

It also frequently happens that scheduled passenger transport undertakings have to bear extraneous charges without receiving sufficient compensation from the State, in which case the community is subsidized by these undertakings. Furthermore the inadequate social conditions of the personnel which are very frequently found in passenger transport also constitute subsidies granted to transport users at the expense of the personnel of these undertakings. For all these reasons an equalization of the conditions of operation of the various transport undertakings becomes necessary wherever existing differences are not caused by natural structural conditions. We consider such equalization as one of the most fundamental prerequisites of the coordination of the various branches of the transport industry.

III. SOCIAL POLICY

Remuneration and social benefits contribute to a considerable item of expenditure in the transport industry. A survey carried out by the International Labour Office showed that on the railways, labour, in relation to the total operating costs of the industry, accounted for up to sixty per cent; in road transport, however, at least so far as Europe is concerned, hardly more than thirty-five per cent. In inland navigation the percentage is perhaps slightly higher than in road transport.

Any proposals for the coordination of transport put forward by the I.T.F. will, for obvious reasons, attach particular importance to the way in which social questions are treated.

It is for that reason that the I.T.F., in close cooperation with its affiliated unions, has been consistently trying within the I.L.O. to promote uniformly agreed conceptions. The Resolution adopted in 1951 which is reproduced in an annex is consequently of fundamental importance for our affiliated unions of inland transport workers.

In times of full employment the transport industry can only rely on a high quality and reliable labour force if the standard of living it provides is at least on a level with that of other industries. Since, however, in addition to risking life and health the transport worker has to bear great personal responsibility and is required inevitably to work irregular hours, as well as at night and on Sundays, an improvement of his social position becomes necessary. Failing this, the safeguarding of the required supply of suitable labour will be prejudiced and the safety of operation endangered. An adequate standard of living of the transport worker, compatible with his obligations must therefore have priority when judging "labour" as an element of cost. The focal point of all efforts towards coordination must be the human being.

The difference in the percentage of costs attributed to labour in the different branches of the transport industry is largely governed by the structure of the industry. The duty to act as common carrier and to maintain operation and the rigid obligations covering timetable and rates to which public enterprises, especially the railways, are subject, lead to a considerably larger establishment. Very often such personnel come under a legal or otherwise politically influenced statute which imposes greater social charges on enterprises than exist in private industry. The greatest possible equalization of social conditions of the personnel in the different branches of the transport industry is one of the most important prerequisites of fair competition. This equalization can only be achieved by bringing the less favourable conditions into line with the more progressive ones. Working conditions which help to maintain the physical and mental labour potential of the worker are of decisive importance to safety in transport. Experience has shown that overtiring, worry and strain are often the cause of accidents. This evil cannot be overcome if better working conditions are merely embodied in laws and agreements. It is equally important to ensure their strict observance by means of an adequate supervision. This must also be extended to such enterprises where the personnel consists exclusively of the owner and members of his family, as in the case of private owners in inland navigation and small road haulage undertakings.

A further aspect of the inequality of working conditions to be taken into consideration is the exploitation of the worker for work of a secondary nature or for additional duties. We find this in one-man operated public transport, in road haulage, where the driver is also called upon to load or unload, also in civil aviation, where the pilot is being given ever-increasing duties. In these cases, an intensity of work is often reached which makes it impossible for those responsible for the safety of operation to carry out their duties during a prolonged period of time without showing signs of fatigue. The only remedy is a corresponding reduction of working time and of the individual shift.

The European transport workers' unions affiliated to the I.T.F. consider the following programme as the basis of the social policy of a coordinated transport industry:

1. Remuneration

Salaries and wages should be compatible with the nature of the work, responsibilities and special hardships of the transport industry. They should above all fulfil a directing function by being of a sufficiently high level in order to attract workers to the transport industry. Other special requirements apart, the need for safety in operation alone should warrant a wage level which would also improve the possibilities of recruiting the most suitable staff.

A further aim should be a guaranteed minimum wage. Reasonable allowances in an appropriate form should be paid for special duties, e.g. night work, work over week-ends and on public holidays.

2. Working hours

In view of technical advances, increased intensity of work and the excessive physical and mental strain on the worker, a reduction of working hours should be aimed at; it should not entail any reduction of the worker's real income.

Working hours, rest periods, rest days and adequate annual vacation with pay are to be adapted to the particular circumstances of the transport industry and the responsibility of the personnel. In particular, the working conditions of operating staff should take into account every aspect of safety in operation.

3. Old age, invalidity and dependants' pensions

In the modern welfare state every individual is entitled to social security in case of invalidity or after reaching the age limit. In case of death, dependants should be assured of an appropriate standard of living. The transport workers' unions insist that appropriate statutory measures are introduced.

As long as no general insurance covering old age, invalidity and dependants exists the transport undertakings should make provisions for industrial insurance schemes with adequate benefits.

Wherever universal insurance schemes exist but their benefits do not suffice in order to safeguard the social security of the workers, industrial institutions shall be created complementary to the general insurance scheme, mainly with a view to covering premature invalidity.

Due consideration should be given to the dangers and physical strain involved in the different types of work when fixing the age limit.

These industrial complementary insurance schemes shall be jointly administered and subject to appropriate supervision.

4. Other social institutions

The social security of the workers and their dependants in case of illness, unemployment and temporary invalidity must be safeguarded in a satisfactory manner. No difference should be made between the benefits to which wage and/or salary-earners are entitled.

In case of temporary invalidity or redundancy owing to technical or organizational changes, other employment should—after retraining, if necessary—be found for the worker concerned which will assure him of an equivalent income.

Employers should provide rest rooms, canteens and accommodation in keeping with the requirements of modern industrial hygiene.

Transport undertakings should make provisions for the systematic vocational training of their staff in close cooperation with the trade unions. They should, in addition, sponsor to the best of their ability sport, physical training and general education. Where similar institutions exist or where they are to be created in order to engender a good working atmosphere and develop the team spirit they should be jointly administered.

In inland navigation permanent arrangements should be made to safeguard the general welfare and health of dependants of the personnel. In this connection particular attention should also be devoted to the proper education of the children.

——o——

The application in all undertakings of this framework of principles can safeguard an adequate supply of personnel, as far as both quality and quantity are concerned. In this manner the transport industry will be able to make its due contribution to the life and work of the people and the nations. At the same time competition between the forms of transport will be deprived of the factor primarily responsible for the distortion of the elements of such competition. It will no longer be possible for an undertaking to offer cheaper transport, at the expense of the social conditions of its workers, than a progressive undertaking. At the same time the threat to the social standard of the most favourably placed workers which they have achieved after decades of trade union endeavour will be removed.

IV. POLICY GOVERNING THE GRANTING OF CONCESSIONS AND PERMITS AND CAPACITY CONTROL

A. General Principles

1. Objectives

As already explained, the purpose of coordination in transport consists in the opinion of the unions affiliated to the I.T.F. in the satisfaction of the transport requirements of the national economy at the lowest possible cost. This objective may only be reached if the capacity of the means of transport available to the community is adapted to transport requirements. This means that expensive surplus capacity should be avoided and, on the other hand, the anticipated future development of the demand as well as possible peak demands should be taken into account.

In our opinion, the only efficient method by which this objective may be reached consists in the introduction of concessions and permits, the latter within the framework of an adaptable system of restriction of capacity. In this connection it would be indispensable to impose the obligation to acquire a permit on commercial and industrial undertakings which carry out transport on own account. Such permits may, according to circumstances, be restricted to one commodity or be related to several commodities.

2. Definitions

The term "concession" as used in this connection implies a contract based on public law, concluded between the general public and a transport undertaking. By virtue of this contract an obligation is imposed on the undertaking concerned to maintain a public transport service between two given points for the duration of the contract. The rights and obligations of the undertaking, particularly regarding the operation of the service and the financial conditions thereof are described in the contract.

A "permit" allows an undertaking, in exchange for a number of obligations, to carry out transport operations within a certain period of time. The rights and duties implied in a permit are less extensive than those which are based on a concession.

The term "concession" as used in this context largely corresponds to the legal conception of the "Certificate of convenience and necessity" used in the U.S.A. whereas the permit as defined here can best be compared with the "Permit" issued by the U.S. Interstate Commerce Commission.

3. The granting of concessions

Concessions shall be granted to undertakings which maintain regular transport services (scheduled passenger and/or goods transport) disregarding the fact whether these transports are carried out by rail, road, or inland waterways. The transport undertaking which is granted a concession will therefore in principle be placed in a privileged position so far as conditions within the particular branch of the transport industry are concerned. The concession makes it obligatory for the undertaking to provide public transport services (obligation to carry and to maintain operation) at rates which are embodied in a rates structure approved by the authorities and which are duly publicized. It will furthermore contain details of the conditions of service (duration and frequency of services, type of vehicle to be used, etc.).

In accordance with the principles of competition which we advocate several concessions may be granted in goods transport for one and the same line, in accordance with the prerequisites mentioned in paragraph B, provided that no ruinous competition is caused thereby.

The granting of a concession does not preclude the activities within the area covered by such concession of an undertaking which only holds a permit. The number of consignors to be served by carriers holding permits per journey or period of time should be subject to special restrictions in order to prevent any undermining of the volume of traffic calculated to be handled by the carrier holding a concession or the "skimming" of the most profitable part of such traffic.

4. The granting of permits

A permit is required by any undertaking which engages in non-scheduled transport operations (passenger or goods) which are either not covered by a concession or occur additionally, as outlined in the preceding paragraph. The permit contains details of the area in which the undertaking may operate, the types of traffic and capacity approved (number of seats and/or total loading capacity, etc.). It also contains the obligation of the undertaking to adhere to the approved rates.

B. Principles Applicable to Scheduled Goods Transport

1. Prerequisites of the granting of concessions

A concession may be granted if:

(a) there is an economic demand for the envisaged service;

(b) all interested undertakings have been given an opportunity to apply by means of a public announcement;

(c) the economic and financial efficiency of the undertaking and/or its economic self-sufficiency can be relied upon;

(d) the applicant and/or his deputy or authorized agent enjoys a good reputation, is reliable from a social point of view, and has the necessary technical, commercial and professional knowledge and experience.

2. Procedure for the granting of concessions

(a) Before granting a concession, the competent authority shall grant a hearing to the joint body of the branch of the transport industry concerned.

(b) Applications by public and private transport undertakings shall be considered on a basis of equality, provided there are no reasons connected with the regulating of traffic which would justify a preferential treatment of undertakings in public ownership.

(c) The concession which is granted shall not impede any change-over to a different technique of operation.

3. Prerequisites of the withdrawal of the concession

The concession shall be withdrawn if

(a) the applicant has knowingly given false information relating to the granting of the concession or is guilty of gross negligence in this connection;

(b) the owner no longer fulfils the conditions on which the granting of such concession was based;

(c) the holder is guilty of repeated and severe infringement of his public obligations, particularly in relation to rates, and the stipulations of social and labour legislation.

4. Validity and transferability

A concession shall be valid for the duration of the whole period of depreciation. It should be renewed provided there are no serious reasons why this should not be done.

The concession shall be transferable by consent of the competent authorities.

5. Special conditions for the granting of concessions in rail transport

Applications for concessions by new railway companies should normally not be granted. If application is made for the construction of a new line for which there is an economic demand and the remunerativeness of which is not in doubt, consideration should be given to an existing railway.

C. Principles Applicable to Non-Scheduled Goods Transport

1. Prerequisites for the granting of permits

A permit may be granted if:

(a) there are no objections to the introduction of the envisaged service from the point of view of the national economy;

(b) the economic efficiency of the undertaking concerned can be relied upon, or, in the case of transport for own account, if its economic usefulness has been confirmed;

(c) the applicant enjoys a good reputation, is reliable from a social point of view and possesses the necessary technical, commercial, and professional knowledge and experience.

2. Procedure for the granting of permits

(a) Before granting a permit the competent authority shall grant a hearing to the joint body of the branch of the transport industry concerned.

(b) The permit which is granted shall not preclude any possibility of a changeover to another technique of operation.

3. Withdrawal of permits

The permit shall be withdrawn if:

(a) the applicant has knowingly given false information relating to the granting of the permit or is guilty of gross negligence in this connection;

(b) the holder no longer fulfils the conditions on which the granting of such licence was based;

(c) the holder is guilty of repeated and severe contraventions of his public obligations, particularly in relation to tariffs, and the stipulations of social and labour legislation;

(d) it becomes apparent that the holder of the permit encroaches to a considerable extent on the field covered by a concession.

4. Validity and transferability

The permit expires at the end of a period of depreciation to be fixed by the authorities and a new application should be submitted.

Transfers of permits should be limited to a few exceptional cases, e.g., in the case of invalidity, retirement or death of the holder.

5. Prerequisites for the control of transport capacity

(a) In order to avoid surplus capacities in transport, the granting of permits in goods transport should be connected with a restriction of the carrying capacity of the vehicles by fixing a total number of vehicles as a ceiling; no further permits should be granted once this ceiling is reached.

(b) This total figure shall be adapted to changes in the demand for transport at certain longer intervals. In times of an economic recession this may also imply the refusal to renew and/or transfer permits and a withdrawal of vehicles from service on a quota basis.

(c) The restriction should apply to the loading capacity, i.e. the capacity of the vehicles (lorries and trailers) for which the permit is granted.

(d) The volume and the regional distribution of the total number of vehicles may be based on the following factors: transport requirements, average utilization of the available loading capacity of the various branches of the transport industry, density of the existing network of roads, demographic and economic structure of the area, total volume of traffic to be handled as well as the rate structure.

D. Principles Applicable to Passenger Transport

1. The principles enumerated under B.1—5 and C.1—5 largely apply also to passenger transport.

2. The actual service which is offered is of particular importance in passenger transport. This means primarily punctuality and adherence to regular schedules and also comfort, accessibility of destinations and safety of operation. Particular importance will consequently be attached to these points so far as passenger transport is concerned.

3. In addition, as already explained, a more far-reaching restriction of competition will be necessary in passenger transport. Competition along one and the same line between scheduled and non-scheduled traffic does not serve any useful purpose. Such competition would not only involve the risk of a disastrous undercutting of fares but also lead to available capacities remaining unused as well as to a lower level of operational safety and punctuality, and, finally, to considerable financial losses. The possibility of these risks being passed on to the transport user cannot be altogether excluded.

4. In connection with the granting of concessions definite provisions should be made for the position of stations, possibilities of transfer and connections with other lines and branches of the transport industry. If necessary, feeder services should be arranged for the benefit of transport users.

5. A policy of coordination of the branches of the transport industry could usefully be implemented on a regional basis. The granting of regional concessions appears to be desirable.

6. In the case of bus undertakings which maintain services wholly or partly in parallel with an already existing line or which merely constitute an extension of already existing lines, the concession should be granted to the undertaking which already operates those lines. Any running in parallel would thereby be prevented so far as possible.

7. No permit should be required for passenger or mixed motor vehicles with a maximum of eight seats. This stipulation shall not apply to taxi services, which shall still require permits.

8. The process of industrial concentration of undertakings of the scheduled passenger transport should be encouraged from the point of view of social policy and in the interest of industrial efficiency.

—o—

In order to supply a criterion for the evaluation of our proposals concerning the policy covering concessions, permits and capacities, we wish to point out that they imply comparatively far-reaching interventions by the State. On the other hand the implementation of our suggestions offers considerable advantages in comparison with other regulating measures (e.g. taxation), because it allows for an adaptation to individual situations and particular industrial contingencies. Conversely, overall measures are not flexible and of necessity also affect cases where special considerations might be justified in view of the particular circumstances involved.

These reasons are also borne out by the fact that even countries with distinctly liberal conceptions, such as the U.S.A., apply a comparatively rigid policy on concessions and permits in the transport industry. One reason, and by no means the least important, is the fact that the only alternative would be very sweeping controls unless one wishes even to go as far as to envisage a one-sided protection of public transport undertakings which are subject to the obligations of a public service. The implementation of our above-mentioned proposals would consequently appear useful precisely in those instances where any gradual elimination of the clear-cut division between the industrial tasks of the branches of the transport industry and their duties as a public service is to be prevented.

This is of even greater importance in view of the fact that the interests of those employed in the transport industry can only be safeguarded if appropriate legislative measures enable the individual carriers to fulfil the social requirements. We consequently wholeheartedly agree with the Resolution of the I.L.O. to which we have already referred (see Annex) according to which a close link between the granting of permits and licences on the one hand and the fulfilment of social obligations on the other should be one of the most important aspects of the procedure governing the granting or retention of such concessions and licences.

It should, incidentally, be pointed out that similar far-reaching measures which are intended to take into account, and by no means in the last resort, the conception of a gradual industrial concentration, particularly in scheduled transport, may only become fully effective after a prolonged period of transition.

V. TRANSPORT RATE POLICY

A. Public Transport Services and Competition between the Branches of the Transport Industry

In the opinion of the I.T.F. transport within a regulated competition should in principle be subject to the general laws of supply and demand. As already explained we believe that deviations from this rule are only appropriate in scheduled passenger transport because overall considerations of social and economic policy as well as of safety of traffic justify planning measures which are more far-reaching than in the case of the remaining forms of transport. Reduced fares in commuters' and school-traffic in support of a sound demographic policy of the State should not lead to a subsidy being granted to these transport users via the means of passenger transport. In so far as rates have to be calculated below the effective prime costs, the loss of receipts caused thereby should be compensated. A compensation from public funds would in this instance be preferable to a purely internal compensation within the framework of the structure of passenger fares. The resulting loss of income may, however, in both instances only be restricted to tolerable limits if the particular branch of the transport industry is protected from competition by fairly extensive measures. Conditions in goods transport on the other hand, are largely different.

We can only agree with the objective of decentralizing industries as far as possible in the interest of economic policy in so far as it does not entail a reduction of productivity. We furthermore believe that this objective may more easily be reached by direct subsidies to remoter industrial areas or by tax relief for the industries concerned rather than by the outdated method of subsidies via the goods tariffs. The history of transport has proved that an excessively close link-up between the planning of location of industries and tariff policy has frequently caused a distortion of the natural elements of production within a given economic area and has been one of the more deep-rooted causes of the unbalanced development of the transport industry.

It is, on the other hand, true that the tariff policy of the railways which aims at service to the community and which has been applied for many decades has largely contributed to the now traditional geographical position of industrial undertakings within Europe. An immediate abandonment of this tariff policy—if all other factors remained unchanged—may consequently lead to economic breakdowns within the remoter industrial regions and cause severe internal disturbances within the national economy which might jeopardize the continuity of the entire process of production. We believe, however, that all essential elements of goods tariffs established in order to satisfy the needs of the community can be maintained also within a State-regulated competition based on price and production and that there is consequently no reason to fear any more far-reaching repercussions of competition in goods transport on the traditional structure of the location of European industries.

The rate policy which is applied in the interest of service to the community is characterized primarily by higher charges on more highly-valued goods for the benefit of bulk goods of lower value by means of the *ad valorem* scale, cheaper transport over longer distances by means of the distance scale and the absence of a strictly cost-related differentiation within the secondary tariff categories as compared with the main tariff category (wagon-loads) by means of the quantitative scale. The application of equal rates within a given area, the obligation to carry, to maintain operation and to publicize tariffs are also often considered as basic principles of a public transport service.

B. Competition and Rate Policy

The original industrial purpose of the *ad valorem* scale consisted in charging goods of higher value with the larger part of overheads and to increase the transportability of bulk goods by applying to them rates which only cover a small fraction of their share of overheads. These differences in the classification of goods of higher and lower value can admittedly not be justified on grounds of costs. This type of rate discrimination could consequently only be successfully applied as long as the railways were in a position of undisputed monopoly. With the advent of an efficient competitor, namely motorized goods transport, and the under-cutting of rates practised by that branch of the transport industry considerable quantities of highly-valued goods have deserted the railways for road haulage. In countries where the latter is subject to rates linked with those of the railways, it was obvious that the road haulage undertakings showed preference for the transport of goods of high value on the very busy main lines. In both instances the result has been the undermining of the principle underlying the *ad valorem* scale.

From the economic point of view, the *ad valorem* scale was intended to compensate for the geographical disadvantages of industries in remoter regions by means of low rates for bulk goods, particularly for fuel and raw materials and by means of high rates for manufactured products, and also to counteract the industrial tendency towards excessive concentration. In reality, however, the transfer of higher-value goods from rail to road and also the lowering of the upper tariff classes combined with a raising of those for bulk goods which the railways introduced on competitive grounds has for quite some time caused the *ad valorem* scale largely to lose its functional value. Furthermore, the continuous lagging of tariffs behind the general level of prices has already led to a reduction of the share of transport costs in the sales prices of goods. Other factors which affect the location of industries, such as taxation, labour supply, existence of feeder and ancillary industries, proximity of consumers, and vicinity of the coast are nowadays of decisive importance for the choice of sites of industries. A possibly even more accentuated levelling of the *ad valorem* scale caused by competition between the branches of the transport industry would consequently merely accelerate a development which is already taking place without causing any danger of far-reaching repercussions to the existing structure of the location of industries.

Conversely, competition between the branches of the transport industry will result in a trend towards a more accentuated reduction of rates for medium and longer distances which is bound to benefit the remoter areas. It is an undeniable fact that the main advantage of the railways consists in the tapering off of costs per unit of production (ton/kilometre) with increasing distances. At the present time this advantage, so far as costs are concerned, is insufficiently translated into railway rates. A competition between the branches of the transport industry based on price and production will, however, compel the railways to adapt their tariffs to a greater extent than they have done up to now to the average costs which become progressively lower as distances increase. Due to the reduced costs at which the railways operate over greater distances they enjoy a strong competitive position in this type of traffic. This enables them in turn to introduce a differentiation within the distance scale for the benefit of the lower freight rate categories for the benefit of the remoter areas which cannot be justified on grounds of costs. It might thereby be possible to better assist industries situated far from the sources of supply of raw materials in obtaining raw materials and semi-manufactured products and better protect their home markets than by the uniform tapering-off of rates which is the general rule nowadays.

The conscious refusal of the railways to base a quantitative scale differentiated in accordance with the actual weight of the wagon-loads on their actual costs is undoubtedly in keeping with the concept of a public service. Industrial production in remoter districts is mainly the responsibility of medium-sized and smaller undertakings which frequently are not in a position to despatch full wagon-loads (which count as such according to their weight) and consequently to avail themselves of more favourable rates. Very likely it will be precisely the competition between the branches of the transport industry which will contribute to the differential freights for smaller consignments remaining within certain limits for the benefit of the trading public in remoter districts and to being calculated far below the actual costs. This development would by no means interfere with the intrinsic interests of the transport undertakings so long as the rates applied to smaller quantities exceed the full amount of the variable costs of such operations. In the case of single consignments up to a total weight of twenty tons (small quantities scale), the competition between the branches of the transport industry will largely lead to a parity of their rates. Such parity would certainly be cost-related since modern research has shown that the tapering off of costs in relation to the weight of the quantities despatched is practically the same for railways and road haulage. The railways are, however, in a considerably better position whenever complete trains—and wagon loads—are despatched.

Furthermore, competition between the branches of the transport industry will compel the railways to express this advantage by means of special rates between traffic centres whenever longer distances are involved. This will cause an extensive return of the large flow of bulk goods from the roads to the railways.

The argument against the introduction of special rates between traffic centres of the railways has frequently been that such rates would intensify the industrial tendency towards excessive concentration and exert a negative influence on the remoter areas. We are of the opinion that rather the contrary is the case. For competitive reasons, the railways will be very interested in establishing connections between industries in remoter areas and routes between traffic centres by means of special distributive through-rates. This tendency will be facilitated by the fact that a considerable number of rail junctions are situated in remoter areas. The introduction of special rates between rail-junctions would, however, also in many instances counteract the one-sided tendencies of industry to concentrate alongside inland waterways and thereby lead to a more intensive decentralization of industries. The assertion that special rates between traffic centres constitute an infringement of the principle of geographical equalization of rates is admittedly theoretically correct. They do, however, cause no negative repercussions on the industries in remoter areas if these industries are enabled to participate in the advantages of transport between traffic centres by means of special through-rates.

It is worth mentioning in this connection that negative repercussions on inland navigation caused by an extension of the system of special rates between traffic centres are by no means unavoidable. The introduction of such rates which would reflect the cost position of the railways in a tariff structure applicable to the entire network should indeed allow for a certain approach of the competitive rates hitherto applied to those obtaining in inland navigation.

It is, however, contended not only in connection with special rates between traffic centres but also quite generally with reference to the demand for greater flexibility of the rate-structures of the various branches of the transport industry that they would nullify the geographical equalization of rates. Apart from the fact that this principle inherent in public service has already been largely undermined in many instances, the introduction of maximum and minimum rates would also compel the individual competing carriers to approach their freight rates as far as possible to the level of costs within the framework of the approved tariffs. They will always have to take into account the potential danger of a restriction of the market by transport on own account, which may be introduced perhaps on a cooperative basis, even in thinly-populated areas with few industries. From a long-term point of view, it will hardly be possible for the average rates to be calculated very much above the level of average costs of such transport on own account.

Up to the present time the system of fixed rates has generally been preferred. Economic developments and competition, however, call for a rate policy which is flexible and offers an adequate margin for adaptation to changed conditions. In order to enable the branches of the transport industry to choose the appropriate moment for such adaptations, they must be capable of implementation within a comparatively short period.

Fixed rates should therefore be replaced by maximum and minimum rates. Maximum rates are necessary in order to prevent transport undertakings taking advantage of the trading public in case of an economic boom which causes a scarcity of transport capacity. Minimum rates are necessary in view of the fragmentary supply of transport capacity of certain branches of the transport industry because there would otherwise be a danger of unused transport capacities exerting a pressure on freight rates in times of an economic recession with the development of a ruinous competition as the unavoidable result.

In times of economic recession, minimum rates will also be necessary in order to prevent deliberate attempts by large undertakings to ruin smaller competitors. Large undertakings with considerable financial resources would be able to carry out such a policy by means of their internal balance of operating costs for some time by means of freight rates which need not even cover the variable costs, let alone the full average costs.

Similarly, the principle of compulsory publication of tariffs as well as the obligation to carry and to maintain operation can be applied equally well within a system of flexible rates. It may be presumed that these principles inherent in the conception of a public service will be applied even more comprehensively within a planned competition because regulated market conditions will make it compulsory to equally distribute these charges among all transport undertakings. The extension of the obligation to carry to tramp traffic will admittedly meet considerable difficulties, by virtue of the particular function of this traffic in relation to the national economy. A solution of this problem should nevertheless be attempted. Experience has shown that tramp traffic will always evade its obligations within the transport industry whenever transports which are not sufficiently lucrative are involved. The consequence is an atmosphere of insecurity in non-scheduled transport which affects the trading public and which may cause an extension of transport on own account. This would, however, involve the danger of a gradual undermining of the economic self-sufficiency of scheduled transport. We consequently consider a rigid organization of non-scheduled transport within road haulage associations and the creation of centres for apportioning freight of their own as desirable. Already existing bodies of this type should be incorporated in such a scheme.

In conclusion it may be said that a regulated competition in goods transport based on price and production does not stand in the way of a public transport service. Whether the collective idea will in future be of any importance at all in goods transport is, however, another matter.

The economic policy of the Western European States aims to an ever-increasing extent at the creation of large-scale common markets which are intended to correspond more to the natural conditions of production in Europe. In the course of this development, the importance of the objective of the national transport policy which consists in the support of remoter areas will in due course be reduced as these regions lose their remote character with the abolition of the economic barriers between countries. The link-up between goods tariffs and the policy governing the choice of industrial sites may consequently gradually be allowed to become more loose and be replaced by a more liberal rate policy which would be better adapted to the requirements of a European common market.

C. Guiding Lines for a Rate Policy in Goods Transport

1. There should be no legally imposed rates parity between the different branches of the transport industry.

The system of fixed rates should be replaced by maximum and minimum rates in order to provide for a greater flexibility of the tariff structure and an adequate margin for a rapid adaptation to the demand at any given time.

Each branch of the transport industry should formulate its proposals for fixing maximum and minimum freight rates by means of joint bodies created for the purpose and submit it to the competent authority entrusted with the examination and approval of rates. Once approved, they should be binding for all undertakings of the branch of the transport industry concerned.

2. It shall be the duty of the authorities entrusted with the approval of tariffs to forestall any tendency towards a ruinous competition when fixing minimum and maximum rates, and to prevent the trading public being taken advantage of in times of a scarcity of transport capacity or in industrially under-developed regions ("reasonable and fair" provisions).

If the branches of the transport industry decided in favour of an *ad valorem* scale, maximum and minimum limits could be fixed separately for each *ad valorem* category. The upper limit of the highest *ad valorem* category shall be calculated in a manner which provides for covering the total costs including fixed costs as well as an adequate margin of profit. The lower limit of the lowest *ad valorem* category should be based on the average of variable costs of the branch of the transport industry concerned.

Should the branches of the transport industry decide to refrain from introducing an *ad valorem* scale and consequently in general only apply maximum and minimum rates, the minimum rates should take into account average costs. The application of this principle is intended to prevent any ruinous competition resulting from the excessive application of minimum rates by any given undertaking.

3. Tariffs shall be deemed as complying with the principle of self-sufficiency of the branches of the transport industry if freight rates of individual undertakings allow for all genuine current operational costs including any necessary re-investments and contributions towards track costs, as well as an appropriate surplus for the payment of capital interest to be covered by the total receipts at the same time safeguarding satisfactory social conditions.

4. After maximum and minimum tariffs or any amendments thereof have been submitted for approval (by the collective body of the branch of the transport industry concerned) the competent authority shall inform that body of its decision at the latest within a very limited period (e.g. two months). If the application is refused the reasons for such refusal must be adequately explained and the applicant shall be informed of the tariff limits which would be acceptable to the competent authority.

5. Maximum and minimum tariffs shall be published immediately they are approved. They shall remain in force for at least one year, provided there are no compelling reasons such as extensive changes of the economic prerequisites which might justify an earlier termination of the period of validity.

6. The individual undertakings of one and the same branch of the transport industry may fix their tariffs within the maximum and minimum limits at their own discretion. These decisions will unavoidably influence areas of different size in accordance with the magnitude and field of activity of the undertaking. Larger undertakings, above all the railways, should therefore be entitled to adapt themselves to varying local conditions. Tariffs fixed in this manner should, so far as scheduled transport is concerned, be no longer subject to official approval but still subject to the obligation of being publicized.

7. In the case of special rates imposed by the State the transport undertakings which apply them shall be compensated for the loss of revenue incurred.

8. Apart from the authorities entrusted with the approval of tariffs, special authorities for the supervision of their application shall be created. The latter shall ensure that the approved maximum and minimum tariffs are not exceeded or undercut and that the tariff differentiations within the limits of the maximum and minimum tariffs do not lead to any discriminations against consignors in a comparable position. Stringent measures shall be taken in case of any infringements of the approved tariffs and the stipulations concerning discrimination.

9. If several undertakings of one and the same branch of the transport industry simultaneously show deficits the authorities entrusted with the supervision of tariffs shall be entitled to investigate in consultation with such undertakings whether and to what extent these deficits are attributable to the tariff policy. If necessary the appropriate amend-

ments should be carried out, with due consideration of the competitive position. Similar measures should be adopted if all branches of the transport industry simultaneously show noticeable deficits.

D. Detailed Problems of Rate Policy in Goods Transport

1. Ad valorem scale

The orthodox basic principles of the freight rates policy of the railways which found expression in the *ad valorem* scale are nowadays no longer of decisive importance. Any differentiation which involves higher charges on goods of higher value is bound to lose its importance with increasing competition between the branches of the transport industry.

The commercial purpose of the *ad valorem* scale has furthermore been undermined by the railways themselves. On the one hand, goods with distinctly bulky character which remain with the railways for various technical reasons (loadability, lack of return loads in traffic radiating from a centre, etc.) are carried at rates which are excessively high in relation to the level of costs; on the other hand, attempts are frequently made to attract greater quantities of more expensive goods (e.g. mixed goods) by lowering rates to a level which barely covers the variable costs.

Quite apart from similar developments, the question arises, however within the transport industry whether in times of a physical shortage of important raw materials (e.g. coal) the transport of bulk goods ought to be encouraged at all by means of lower freight rates. We believe that it would be more appropriate to apply the laws of supply and demand for the benefit of the transport industry within the existing possibilities of determining freight rates.

Due to these commercial reasons and the economic considerations contained in the second paragraph of this chapter the importance of the *ad valorem* scale has been considerably reduced. This is also demonstrated by the tendency of the railways to compress the *ad valorem* scale for reasons of competition. We nevertheless believe that its complete abolition would not serve any useful end. Goods in higher tariff classes which have remained with the railways are for special reasons (better loading facilities, large size in relation to weight, dimensions, time factor, etc.) in many instances "railway-biased" goods which cannot easily switch over to road haulage. An exaggerated compression of the *ad valorem* scale, let alone its abolition, would result in granting the consignors of these goods a freight discount, whereas the railways would not always be in a position to make good these losses through an increased volume of traffic. The railways would, on the contrary, risk losing bulk goods if they became subject to relatively higher freight rates.

An extension of the *ad valorem* scale would, on the other hand, not be in keeping with the competitive position.

2. Freight rates tapering with distances and rates between traffic centres

We have already explained that the main advantage of the railways so far as costs are concerned consists in the tapering off of costs per unit of production (ton/kilometre), an advantage which becomes more accentuated as distances increase. In the present rates this advantage is only insufficiently expressed, particularly so far as medium distances are concerned. On the other hand, short hauls are frequently carried out at a rate which is sub-normal in relation to the high terminal charges.

Whereas, therefore, the advantages offered by the tapering-off of costs ought to be expressed in the rates applicable to medium and long distances, the rates for short-distance traffic should be determined strictly on a cost basis. Reduced freight rates within a short distance zone would only be useful for complete trains and groups of wagons because short hauls in large quantities are a paying proposition. Similar reduced rates would also be in the interest of inland navigation which would otherwise, if short distance rates were increased, be forced to adapt itself to a considerably more expensive traffic involving transshipment.

Over longer distances between large centres of rail traffic the railways are in a particularly favourable cost position. This fact ought to be reflected in special rates between such centres by which means dense traffic streams could be attracted to the railways.

In inland navigation considerably reduced rates are frequently charged over longer distances (up to fifty per cent). The network of inland waterways, however, is in most countries based on traffic centres. If the idea of special rates between such centres is now applied equally to "dry" centres of traffic it amounts to no more than the long overdue adaptation of inland transport to existing conditions.

At the same time, through-rates between centres would also benefit consignors and consignees in remoter districts, namely in so far as transport follows routes between traffic centres for part of the total distance. Since, furthermore, numerous "dry" traffic centres are situated in outlying areas, the scheme would cause rather favourable economic repercussions for these areas. So far as railways are used in this connection the tendency to use road haulage over the total distance would be reduced; this tendency is already noticeable wherever feeder services have to use the roads. A rapid adaptation of the railways to the development of mixed traffic (pickaback) appears to be of particular importance in this connection.

3. Quantitative scale

The differences in freight rates between small consignments (e.g. five to ten tons) and complete wagon loads (fifteen to twenty tons) by no means correspond to the actual position so far as costs are concerned, because small consignments very often cause considerably higher costs in comparison with full wagon loads. This is essentially due to the fact that particularly in the case of comparatively light but bulky goods and goods which do not load well only a fraction of the actual loading capacity can be utilized.

We therefore advocate in principle higher rates for the secondary categories, although, because of considerations of an overall transport policy, it will admittedly never be possible to take into account the entire difference in costs. Consequently variations within the scale for small quantities will also hardly be conducive to an increased profitability of the different branches of the transport industry. This objective could only be reached if a comparatively greater increase in secondary freight rates led to a better utilisation of the capacity of goods wagons. There are, however, certain limits. The relatively lower degree of utilisation of the loading capacity of railway trucks, compared with that of lorries (fifty per cent compared with seventy to seventy-five per cent), is mainly due to the incidence of traffic radiating from a given centre, where loads of raw material in one direction are only rarely compensated by full return loads.

Another aspect to be taken into consideration in connection with an increase in secondary freight rates is the fact that even part loads which, when accumulated, result in full loads or help to avoid empty runs may be of some commercial importance to the branches of the transport industry. The differences between the cost of the carriage of small consignments and full loads are roughly the same in both road haulage and rail transport.

If any amendments of the rate structure of the railways in favour of the application of special rates between traffic centres were envisaged, the introduction of special rebates for complete trains and groups of wagons would admittedly be of reduced importance. The fact that these rates offer a considerable advantage to the railways from the point of view of costs, should nevertheless not be overlooked. The additional introduction of such rates for larger and very large quantities may therefore appear appropriate provided the above-mentioned advantage of the railways is not already counter-balanced by special rates between traffic centres and the extension of the distance scale for the benefit of the transport users along the lines concerned.

Wherever competitive rates in relation to inland navigation are being applied, the introduction of the large quantities scale would make such competitive rates largely superfluous.

Summarizing the above considerations, one could therefore say that the tariff reforms which we have described would hardly exert an undue influence on the competitive position of inland navigation.

E. Special Aspects of Goods Transport

Container traffic and combined transport (*pickaback*) should be subject to a uniform rate for all branches of the transport industry policy. The rates to be applied should aim at the promotion of further development and unification of these transport methods and consequently lead to a natural cooperation between the different branches of the transport industry.

The transport of *parcels of mixed goods* to less frequented areas causes disproportionately high costs. An adaptation of delivery times to similar special conditions as well as improvements of the system of bulk collections and deliveries of forwarding agents would be of considerable assistance.

Auxiliary charges ought not to be levied in a bureaucratic manner as if they constituted a special compensation for some administrative service rendered. Their use should rather aim at the prevention of abuses; particularly demurrages fulfil an important function by preventing rolling-stock remaining idle. They ought to be more variable according to the season in order to counteract a seasonal slowing down of the turnover of wagons.

Secondary lines ought to cover their specific costs. As an alternative, the granting of subsidies on a regional basis might be taken into consideration.

In the case of *private sidings*, the abolition of charges for them may cause short-term losses of revenue so far as apparently "safe" customers are concerned. From a long-term point of view, the abolition of such charges is likely to further the development of the genuine house-to-house service of the railways which already amounts to a large percentage of wagon loads due to the use of sidings.

Special agreements of the transport undertakings with certain traders could be applied during a transitory period with a view to counteracting transport for own account. They could encompass bulk rates or "agreed charges" as in Great Britain or perhaps quantity- or good-will rebates. They should be subject to official approval but without the obligation to publish them. The competent authority shall refuse its approval if the special tariff (special exceptional rates) does not cover the entire additional cost of the amount of traffic carried. The validity of such rates shall be agreed contractually between transport undertakings and customers. Since certain forms of these agreements cause a continuous expansion of the average transport distance a long validity would in this instance not serve any useful purpose.

F. Detailed Problems of Rate Policy in Passenger Transport

The deficit of nearly all European railways appears to be partly caused by the fact that revenue in passenger transport by no means covers the sum total of costs chargeable to it. Deficits appear particularly in traffic consisting of stopping passenger trains and commuters', students' and other forms of traffic for which special rates are granted because of social considerations. Conversely, that part of the traffic which shows a tendency to lose customers to individual means of transport or civil aviation is in many countries subject to rates which by far exceed costs. The internal equalization of costs which is thereby obtained may appear appropriate from the point of view of the national economy as a whole. In the majority of cases, it does, however, not suffice to provide for economic self-sufficiency of the branch of the transport industry concerned.

Even if as we suggest, far-reaching regulating measures were adopted in order to provide for a certain protection of scheduled passenger transport from competition, the desertion by part of the users in favour of individual transport cannot be halted. This development should be taken into account when establishing the rate policy for passenger traffic by means of fast and super-fast trains. On the other hand, the loss of revenue occasioned by the application of such special rates must be compensated by state grants. An internal cost equalization may no longer be adequate under modern conditions.

Passenger fares on secondary lines should be calculated in such a manner that they facilitate the closing of lines which do not pay and also facilitate the transfer of a reduced but widely distributed volume of traffic to the road. It will, however, be necessary, when judging the utility of closing a line, to recognise the fact that certain joint expenses (for instance, for maintenance and renewal of tracks) cannot be saved and would after the closing of a line to passenger traffic solely constitute a charge on goods transport.

VI. INVESTMENT POLICY

A harmonious relation between the development of transport and that of the entire industry is a prerequisite for a smooth functioning of the process of production within the national economy. If the capacity of the transport industry lags behind the general productive capacity, the maximum limit of the national production which could in itself be reached will nevertheless not be obtained. On the other hand, an over-dimensioned transport industry may also exert an unfavourable influence on the national production because it would deprive industry unnecessarily of labour and goods.

A. Investment in Tracks and Vehicles

Special problems arise in connection with the criteria for investments in tracks. In principle the network of tracks should be adapted to requirements governed by the location of industries. Its capacity will depend on the number and technical data of the vehicles, with due consideration to all necessary safety measures.

Investments in construction of roads have to serve two different purposes, namely, firstly, industrial production by providing space for the collective movement of goods and passengers and for the vehicles used in transport for own account. Secondly, they are intended to meet the requirements of individual transport, i.e., to satisfy not a requirement of production but simply a consumer demand. In the majority of countries the road-space taken up by cars, motor-cycles, mopeds and bicycles has increased on a larger scale than that used by commercial transport. As a consequence, the construction of roads can nowadays no longer be planned primarily with a view to satisfying the requirements of industrial production but the demand of a large proportion of the population for better facilities for individual transport must also be taken into account. In view of this development a factor of insecurity begins to exert an influence on investments planned in road construction. The attribution of their share in the cost of tracks to the various branches of the transport industry is thereby made considerably more difficult and any timing of investments purely in accordance with the requirements of industrial production becomes impossible.

Despite these difficulties in connection with the construction of roads all track investments should satisfy the requirements of overall economic self-sufficiency. This means that each transport undertaking should defray, from its own means, all costs of tracks which it uses. In the case of investments in projects competing with already existing tracks it should, in addition, be necessary for the transport service provided by the new installations to cost less per unit of transport carried than by the already existing facilities. When making such a comparison between the total costs of the old installations and those of the new installations to be constructed, the capital losses to the national economy incurred by the closing down of the old but not yet completely written off installations and the interest on capital incumbent on their remaining value must be assessed and added to the total costs of the new installations to be constructed. In principle, investments in vehicles should only be approved if justified by the quantitative transport requirements, to be assessed on the bais of past demand and the demand anticipated in the future. The transport undertakings concerned should be required to prove that these investments are sufficiently covered by their own or outside funds, always taking into consideration the principle of economic self-sufficiency.

The dualism of public and private complementary investments in the transport industry creates a special problem of coordination. Whereas the construction and maintenance of roads and inland waterways generally fall within the field of competence of the State and/or public authorities, the acquisition of vehicles and the determination of their capacity are within certain limits left to the discretion of private undertakings. In the absence of coordination, this division in the planning of complementary investments will cause a disproportional development within the branch of transport concerned and also a distortion of the relations between the various forms of transport.

A policy of investments should consequently aim at a development of the various means of transport and tracks in accordance with the transport requirements of the national economy and the long-term plans of expansion of industry in general, as well as with the objective of increased production. Investments for the purpose of increasing safety of operation in general and protecting those employed in the transport industry should receive priority.

B. Investment and Business Cycle Policy

Investments in tracks are a useful instrument of an anti-cyclical economic policy and consequently ought to be adapted in principle to the long-term requirements of such a policy. The only exceptions from this rule ought to be the requirements of safety to which we have already referred.

We believe that the market mechanism will also have an economically useful influence on the coordination of investments in transport provided the principle of economic self-sufficiency is applied in a more elastic manner. Such an indirect control of investments does, however, not entirely preclude disproportional developments, because long-term investment plans are counter-balanced by short-term fluctuations of revenue, particularly due to cyclical fluctuations. These may cause individual enterprises to take decisions on investments in transport which are undesirable from the point of view of the national economy as a whole.

In times of rising profits, private industry shows a tendency to increase investments, whereas the State and public authorities, due to economic considerations, rather impose certain restrictions on their own activities. Conversely, in times of economic recession, private industry tends to reduce its investments; then, however, increased public investments, particularly in the field of transport, become necessary.

An anti-cyclical policy in road building meets additional difficulties due to the fact that the requirements of individual transport accentuate the above-mentioned attitude of private undertakings to an even larger extent. We consequently consider it necessary for a central body to be entrusted with the task of balancing competing investments in the transport industry from the point of view of their general economic importance and to establish guiding principles for long-term plans of investment for the entire transport industry.

A coordination of the investments of the different branches of the transport industry is vital also in connection with the necessary further development of the container traffic and combined transport. The aim should be to recognize the provision and maintenance of containers to be largely a common responsibility of the different branches of the transport industry and to restrict investments on the part of the trading public in this sector to special cases.

C. Disinvestment and War Damage

A prerequisite for an organic development of the various forms of transport is also a certain willingness to agree to disinvestments in some sectors of the transport industry. In our opinion such disinvestments should always be effected if a partial or total substitution of one branch of the transport industry by another is proved to be advantageous in the interest of the national economy. Economic advantages arising from such substitution may be expected if the sum total of expenditure connected with all factors of production involved in operation and maintenance of the installations which are to be replaced exceeds within a given time the sum total of all elements of production which become necessary for the construction, maintenance and operation of the new installations including the payment of capital interest. Furthermore, the re-employment elsewhere of workers who may become redundant as a result of such changeover investments and, wherever possible in equivalent positions, should be guaranteed.

A particular objective of the investment policy in the transport industry is the opening up of economically underdeveloped and in particular of marginally situated regions. For such opening-up investments, the initiative of the State is required to a greater extent because, due to their long-term character, they would involve too great a risk for any transport undertaking which aims at economic self-sufficiency. The branches of the transport industry entrusted with similar tasks should be compensated by the State.

Similarly, assistance by the State is justified wherever the financial position of an undertaking due to war damage and indirect repercussions of the war stands in the way of modernization which may be urgently required. These undertakings are frequently forced by lack of financial resources to restrict their reconstruction and renewal schemes to the re-establishment of pre-war standards. This means, however, that the transport industry loses the benefit of considerable technical advances. Special public investment grants in order to overcome similar difficulties are not only justified but even necessary in the interest of the national economy as a whole.

VII. TRANSPORT ON OWN ACCOUNT

A. Goods Transport on Own Account

Transport on own account is a factor of some importance in the field of goods transport, side by side with commercial* goods transport. "Transport for own account" means the transport carried out by a physical or juridical person for its own requirements. The vehicles used either belong to such persons or are hired for its exclusive use. It assumes responsibility for the actual transport of goods in its possession or property, or of such goods which form the substance of its productive activity. It is furthermore understood that such transport for own account represents only a complementary or subsidiary activity of the aforementioned physical or juridical person.

It may happen that the goods which represent the substance of the transport for own account do not suffice to provide a full load for the vehicle used for the purpose. In such a case the load is frequently completed by goods belonging to a third party. This is called "false or mixed transport for own account"

In addition to these categories there is also camouflaged transport for own account, which is of some importance in particular in inland navigation. There we find transport undertakings which although juridically independent, are in reality established by an industrial or commercial enterprise for its own purposes and/or under its control. This splitting off of such nominally independent undertakings should not have as a consequence that measures to regulate commercial goods transport and/or transport for own account be circumvented.

From the point of view of transport policy, only transport on own account beyond the municipal areas in which the actual undertakings are located is of vital importance. Delivery service within these areas may be excluded from any controls if it is commercially necessary or forms an indivisible part of the service which customers are entitled to expect from the industry or trade concerned. Should, however, a compulsory application of fixed freight rates also appear desirable within a local zone, regulating would also have to be considered in this field.

We do not wish to deny that genuine transport on own account may not only entail advantages for the undertaking concerned, but also for the national economy. Such advantages occur particularly in the case of less industrialized or less densely populated areas. Similarly, in the case of transport carried out in special vehicles and such transport operations which cannot be carried out rationally by commercial transport, transport on own account may serve a useful purpose. In these cases genuine transport on own account may also be justified on grounds of transport policy.

**Note:* The term "commercial" is understood to apply also to public transport undertakings.

Very often, however, transport on own account carries out operations which could be taken over by a commercial road haulage undertaking without difficulty or disadvantage to the industrial undertaking concerned. Such transport, even if it shows a profit from the point of view of that undertaking, nevertheless impedes the maximum utilization of the transport capacity of the carriers which is available to the public. In road transport the tasks of the operators become more difficult in view of the fact that transport on own account takes up a disproportionally large area of roads.

The tax relief granted to transport on own account and the fact that it is not subject to the obligations imposed on commercial goods transport, have contributed to a continual increase of its capacity. Very often it has indeed extended beyond the actual purposes of the undertaking. This has in no small way contributed to the creation and expansion of false transport on own account. On grounds of transport policy an excessive expansion of transport on own account is bound to cause grave concern, because it creates a serious obstacle to a genuine coordination of transport. On the other hand, transport on own account is in many instances only introduced for the purpose of covering normal transport requirements. In case of additional demand or whenever vehicles employed in transport on own account are laid up, commercial transport must be prepared to fill the gap, whereby considerable additional costs arise.

It would therefore appear to answer the purpose if transport for own account were restricted by introducing a licensing procedure and by controlling its capacity.

We do not consider prohibitive taxation for the purpose of restricting transport for own account as suitable. It would equally affect both that section of the genuine transport for own account which can be of use to the economy as a whole, and the false transport for own account as well as that which could not be justified by transport policy. A special tax could exceptionally be justified if it serves to equalize the conditions of competition. It should then offset the advantages which transport on own account is afforded by the parent company by which it is carried out, and take into account the inconvenience and additional expense caused to commercial transport due to the particular restrictions and obligations to which it is subject.

By means of the licensing procedure and by controlling capacity, it is possible to determine the extent of transport on own account in accordance with industrial requirements as well as that of the national economy and transport policy. The granting of permits for transport on own account should be based on the principles of the relevant policy as already defined elsewhere in this document, with particular attention to the maximum utilization of the capacity of commercial transport.

Permits should only be granted within the framework of a quota established for transport on own account. This quota could, if necessary, be revised periodically, in order to take into account any insufficient supply of capacity of commercial goods transport. The introduction of a system of quotas for transport on own account does, of course, not mean that there would be any obligation to grant permits up to the limit of the established quota.

False transport on own account should be legally prohibited; exceptions should, however, be permissible in special circumstances.

B. Passenger Transport on Own Account

Passenger transport on own account consists of:

(1) transport without direct or indirect remuneration;

(2) passenger transport which is operated exclusively for the account of a physical or juridical person; on condition however, that the vehicles involved belong to the person concerned or are, by agreement, exclusively at the regular disposal of such a person, and that apart from the driver only personnel or trainees are carried, and on the further condition that the transport in question is in direct connection with the activity of the physical or juridical person concerned.

In this instance too, transport for own account, even when it appears advantageous to the individual enterprise, should not be allowed to undermine the transport capacity and remunerativeness of commercial transport. If there is an economically unjustified expansion of capacity then, as in the case of goods transport, restrictions by means of a licensing system within the framework of a quota which would have to be revised, may have to be considered. Otherwise, in cases where the passengers travel free of charge, applications for licences ought to be given favourable consideration. Care should, however, be taken to see that, if possible, no transport of this nature is carried out in lorries.

From the point of view of safety of traffic it is particularly important that the safety regulations applicable to commercial transport be also applied to passenger transport for own account. This means primarily that passenger transport for own account should become subject to the provisions governing construction, interior arrangements and type of operation of the vehicle which are applicable to commercial transport.

For the rest, it appears to be obvious to us in the I.T.F. that personnel employed in transport for own account should, from a social point of view, be placed at least on the same level as their colleagues in commercial transport. In this instance too we fully support the recommendations of the I.L.O., reproduced in the Annex.

———o———

Our proposals concerning transport for own account may on the whole be considered as rather restrictive. It is, however, in our opinion one of those instances where one cannot both have one's cake and eat it.

It would be impossible to provide for an efficient transport industry which serves the interests of the community by means of suitable coordinating measures unless private profiteering which is frequently the predominant motive for the introduction of transport for own account is forced to remain within certain limits. We realize on the other hand that the full effects of the implementation of our proposals will only make themselves felt gradually and after a prolonged period of transition. They also presuppose that commercial transport will make every effort in order to provide the national economy and the trading public with a transport service of high quality at all times.

VIII. COORDINATING AND SUPERVISORY BODIES AND INSPECTION AUTHORITIES

A. Transport Advisory Council

The connections between the general economic policy and the transport policy have already been pointed out. Planning and realization of the policy of coordination are, in principle, part of the duties of the State; the repercussions of this policy are, however, important both for the employers and employees as well as the users in the transport industry. For this reason these parties must be given an opportunity of being consulted in all questions of transport policy.

For this purpose a Transport Advisory Council, under the chairmanship of the Minister of Transport, should be created and should have the following structure:

Each branch of transport to provide an equal number of employees' and employers' representatives who together would constitute half the members of the Council.

A quarter of the members to be provided by the transport users.

The remaining quarter of the members of the Council to consist of experts particularly conversant with the transport problems.

The employees' and employers' representatives to be nominated by their representative organizations.

The Chambers of Commerce, Industry and Agriculture, and where the case arises, the Transport Users' Associations, where they can be described as truly representative, to nominate the transport users' representatives.

The selection and nomination of the experts to be made by the Government.

The establishment of Regional Committees can be envisaged to carry out certain preparatory work and to study regional problems. Their structure should be identical to that of the Transport Advisory Council.

The task of the Transport Advisory Council is to give advice on questions of general transport policy and in particular concerning:

(a) The rate policy,

(b) the programmes of investments in the transport industry,

(c) the granting of concessions and permits,

(d) the capacity quotas allocated within the policy governing permits,

(e) the equalization of competitive working and social conditions within the different branches of the transport industry,

(f) the safety of operation in transport,

(g) the international problems of transport, in particular the fixing of international standards and investments,

(h) all problems submitted to it by the Minister of Transport or the bodies of the different branches of transport, as well as any other transport questions which the Council itself may wish to study.

The final decision on any question examined by the Council rests with the Government. If the Government reaches a decision at variance with the advice, provision should be made for the Government to publish a statement explaining its reasons for not following the advice given by the Council. In these cases the main points of the Council's advice should be made public. The Government may ask the Council to re-examine the particular question before reaching a final decision.

B. Bodies of the Different Branches of the Transport Industry

In order to obtain a collaboration based on mutual trust within the framework of the different branches of the transport industry and to ensure the realization of the coordination policy, a body must be created for each branch of the transport industry, to be composed of representatives of the following groups:

(a) transport undertakings (employers),

(b) personnel of these undertakings,

(c) users,

(d) administration or experts.

The proportional representation of the groups should be the same as within the Transport Advisory Council. Representatives of the trading public and forwarding agents should be included among the users' delegates.

The main tasks of these bodies in their respective branch of the transport industry should be the following:

(a) to propose tariff structures;

(b) to investigate the calculation of costs, profit and loss accounts and statistics, and to suggest methods for their unification;

(c) to make proposals for the organization and operation of the freight agencies and the freight compensation funds;

(d) to consider proposals for the granting of concessions and permits;

(e) to propose investment schemes for their respective branches of the transport industry or to give advice in connection with similar schemes;

(f) to investigate the possibilities of increasing the safety of operation and of the rational organization of the undertakings from a technical, industrial and commercial point of view;

(g) to suggest improvements in the vocational training of the technical and commercial personnel;

(h) to submit to the administration or the Transport Advisory Council any suggestions which they may deem useful.

In so far as it is considered expedient to directly entrust these bodies with questions of coordination beyond their consultative functions the competent authorities should empower them accordingly. In such cases decisions taken by these bodies will be binding for the branch of the transport industry concerned. The Minister of Transport or a transport undertaking can raise an objection to the decisions of these bodies within a given period. If the body in question stands by its decision after the question has been reconsidered, the dispute must be submitted to the Transport Advisory Council.

If it appears useful to study certain problems at regional level first, regional bodies may be constituted in addition to the central bodies. Their structure should be the same as that of the central bodies.

C. Supervision and Inspection

Regulations and measures to be taken within the framework of the coordination policy, as explained in this report, only serve a useful purpose if their application is properly supervised. Special supervisory bodies should therefore be established with the following duties:

(a) To see that the maximum and minimum rates established for the different modes of transport are not exceeded or undercut and that the differentials within their limits do not lead to any discriminations between consignors in comparable circumstances.

(b) In the case of any branch of the transport industry showing a chronic deficit to find out to what extent this deficit may be the result of a wrong rate policy, and if this is the case, to propose appropriate tariff changes.

(c) To supervise the operation of the freight agencies and freight compensation funds.

(d) To supervise the application of the social and labour legislation.

Detailed supervision shall be made possible by the following method:

1. Transport undertakings, freight agencies and freight compensation funds are under the obligation to submit all documents to the supervisory bodies for inspection at their request.
2. The supervisory bodies are authorized to carry out inspections at the premises of all parties concerned in the transport contract or its financial implications.
3. The supervisory bodies should operate control points along the lines and/or roads and at traffic junctions in order to verify documents and see to it that they are properly kept.

The supervisory bodies should be given executive powers in order to impose fines in case of offences or contraventions or, where the law does not permit this procedure, to initiate legal proceedings. For this reason the necessary authority under public law should be vested in these supervisory bodies.

In cases where the maximum and minimum rates have been exceeded or undercut the supervisory bodies should be empowered to order the refund of the difference between the actual charge and the official rate to the undertaking or the consignor, as the case may be. If the undertaking and/or the consignor do not comply with this instruction within a given time the supervisory body should be legally entitled to claim that difference.

In conclusion we wish to point out that the application of the inherent principles of the resolutions on labour inspection in road transport adopted by the 1957 Inland Transport Conference of the I.L.O. to all branches of the transport industry would appear to be eminently suitable to enable an efficient labour inspection to be carried out at reasonable cost.

PART TWO

Special Problems of Coordination

IX. URBAN TRANSPORT

1. Urban development, motorization and traffic space

Similar to the development of modern monetary systems which sponsored the development of towns and cities, the development of industry, commerce, and transport has contributed to a continuous and frequently rapid growth of these towns during the last century. Their outer limitations by city walls were removed and new quarters began to develop around the old nucleus. At the same time a tendency towards the formation of city centres began to appear. The buildings of authorities, administrations, and centres of culture were joined by banks, insurance companies, administrations of large industrial and commercial undertakings, warehouses, and specialized shops of the retail trade. Very little room was left for housing purposes. Industries and housing estates developed in the suburbs. As a consequence the most varied transport requirements arose.

As the demand for transport increased the possibilities of satisfying it also developed. Whereas it was still comparatively easy to take into account this development of modern transport during the period of construction of suburbs and connections between towns and to build roads accordingly, only limited possibilites in this respect existed in the older parts of the towns. Particularly in the valuable and organically developed parts of the towns which constitute the centres of culture commerce and trade and necessitate the daily transport of large numbers of people as well as of a considerable volume of goods, very little space was left for improvements of communications.

The rapid increase in motorization since the second world war has, strangely enough, caused a continuous slowing-down of the pace of traffic in the big cities. The excessive demands on the available road space in the urban areas has reached a stage where traffic is threatened by self-strangulation. This applies particularly to mass transport affected thereby during peak hours when the various bottle-necks caused by individual transport force the average speed down to that of a tortoise. This causes considerable additional expenditure for these transport undertakings and furthermore heavy economic losses in general and finally increased desertion from public means of transport to individual transport. In the last resort it may happen that business undertakings which hitherto have been domiciled in the inner cities decide to move to the suburbs where the lower prices of real estate facilitate large-scale transport and parking arrangements. This would, however, threaten the inner cities with a gradual undermining and devaluation and the concomitant heavy economic losses.

This chaotic development can only be avoided by either adapting traffic to existing possibilities or on the other hand by creating all necessary prerequisites in order to enable traffic to flow without obstacle. So far as the second alternative is concerned, we wish to state immediately that it would imply enormous expenditure which would have to be covered by the community and consequently the taxpayer. Such expenditure is grossly disproportional to any possible profit and may entail very far-reaching repercussions in other fields of official activity such as social policy and education.

A solution of the transport problems is the prerequisite for a healthy development of the urban economies and the private lives of their citizens. It is obvious that a solution of this kind cannot take into account isolated points of view and requirements but only the interest of the entire community.

Possibilities of improvements of traffic conditions in the city centres should at any rate always be examined in close consultation with the town planning authorities. There is little point in adopting measures for the purpose of speeding up traffic in the central areas of cities if on the other hand the construction of enormous sky-scrapers is permitted which require a parking space which exceeds their effective capacity many times whereby those who are responsible for the construction of these buildings contribute little or nothing towards the cost of such parking facilities.

The special conditions prevailing in urban transport call for regulating measures which may not be entirely compatible with our statements in Part One of this study. In the first place, the problem of coordination can in this instance not be considered as an isolated aspect but must be studied from the global point of view of urban economy, urban administration and town planning, which presupposes the creation of the appropriate authorities. Furthermore the purpose served by urban transport calls for measures which involve priority for the vehicles of public transport and essential supplies of goods.

2. Possible regulating measures

Short distance goods traffic in urban areas and the surrounding country consists mainly of three elements: traffic to and from railway stations and ports, commercial deliveries and collections, and transit traffic. The latter need not concern us here in view of the fact that it may be diverted by measures of the traffic police and the construction of by-passes.

So far as the business life of a town is concerned, transport between traders and consignees on the one hand and railway stations and ports on the other are of particular importance. The burden which they impose on urban traffic is even heavier in view of the fact that such transports are frequently carried out by heavy vehicles which—at least so far as local deliveries are concerned—have to stop frequently and consequently cause congestion. It may become necessary not only to limit the capacity of vehicles in the central areas of towns or within certain zones but also in certain circumstances to limit the number of lorries to be admitted. All this would result in the application of some sort of quota system. It would of necessity have to be applied as well to inter-city long-distance traffic on the road. For this purpose forwarding centres would have to be created at a suitable distance from the town centres which would have to be used for trans-shipments of merchandise to the admitted vehicles in the same way as goods stations and ports. In extreme cases it might even be feasible to make one holder of a concession responsible for services within the central areas or certain zones of towns. This arrangement would have to be operated in conjunction with certain strict regulations which would afford priority in the use of roads to passenger traffic during peak hours but which would be applied in a flexible manner to certain streets and periods.

Collections and deliveries within urban areas by means of motor vehicles have shown an extraordinary increase during the past few years. In addition, vehicles of public services and supply undertakings (post, telephone, telegraph, electricity, gas and waterworks, street cleaning, refuse disposal, etc.), are also being used in ever greater numbers. To the extent to which these vehicles circulate within urban areas the introduction of a system of concessions and permits including the allocation of quotas as envisaged in Part I of this study would hardly be feasible. On the other hand we believe that even in this particular instance certain restrictions imposed by the traffic police would be useful which would afford priority to passenger traffic during peak hours and would consequently free the most frequented streets and roads of moving and parked vehicles of that type.

These remarks on possible limitations on goods traffic within the towns are intended to indicate that we attach the greatest importance to a rational transportation of passengers during peak hours. Nobody will deny that this would in many instances very largely reduce the traffic congestion within the towns.

Peak hours only amount altogether to no more than a few hours per day. They are mostly spread over a short period of time before opening and closing hours of shops and businesses. Frequently too additional but less distinct peaks occur towards the middle of the day. During these peak hours the most important task of traffic planning consists in making provisions for the transport of the working population from the suburban areas into the towns and back. This main stream of traffic consists of the users of public means of transport on the one hand and of individual transport on the other. So far as the latter is concerned, the proportional breakdown in motor-cars, motor-cycles, and bicycles varies from one country to the other. Consequently the extent to which the interests of their users will have to be considered will also vary.

Last but by no means least the rights of pedestrians have to be taken into account. Pavements have frequently become inadequate in view of the fact that part of their width has been added to the road. On the other hand pedestrian crossings have been introduced which contribute to a noticeable slowing down of traffic in town centres.

If the streets in the town centres are not of sufficient width and on the other hand there are sound reasons which impede any widening of existing streets or the construction of new ones the question arises as to who has to give way to whom. It is certainly not easy for the competent authorities to find the right answer to that question. They will always endeavour for different reasons to adopt measures which meet with the least opposition among those who are affected by them. Any real remedy, however, can only consist in far-reaching measures which may differ according to the size and structure of a town and the degree of motorization.

We believe that any solution should be based on the principle that all participants in traffic may claim approximately the same rights. The priority of public transport therefore becomes quite obvious. According to recent statistics a bus which holds seventy-five passengers occupies twenty-six square metres of road surface whereas the same number of persons, travelling by private cars with an average of 1.7 occupants per car require approximately three hundred square metres of road space. A comparison between tramways and motor-cars will no doubt be even more eloquent. Whereas nearly three passengers may occupy one square metre within a fully loaded public transport vehicle the driver of a private car requires approximately seven square metres only for himself. This disproportion becomes even more acute if space is taken up by parked vehicles.

All parking of vehicles for longer periods ought to be prohibited in the streets within the town centres which carry heavy traffic. Furthermore, individual means of transport ought not to be allowed to stop anywhere in town centres during peak hours. At the same time, however, arrangements should be made outside the town centres for people who permanently require parking space to enable them to transfer to public means of transport without any undue loss of time. Side streets and parking spaces within the inner areas of towns should only be available for temporary parking and against payment of a special fee. The amount of such fees should be calculated in a manner which would on the one hand efficiently limit the use of individual vehicles for journeys to town centres to essential purposes and on the other hand cover all public expenditure occasioned by parking facilities. We also believe that investment in the construction of multi-storey parking facilities in town centres should be limited to a minimum in view of the fact that facilities of this type encourage individual transport to use the roads in town centres instead of keeping it away from them. These parking facilities could consequently be mainly constructed along the main exit roads. Furthermore narrow streets should only be open to one-way traffic.

Regulating measures of this type could contribute to a considerable speeding up of traffic, mainly in towns up to a certain size. The expenditure which they entail is low in comparison with that caused by widening of streets. They may, however, only serve a useful purpose if the enterprises responsible for public transport are at the same time willing to pay for their regained freedom of movement by the extension and modernization of their transport services.

The reply to the question whether the rail-bound tramways stand in the way of efforts to make traffic flow more easily in the streets of town centres or not depends on numerous factors which cannot be discussed in any greater detail in this study. If one neglects the cost factor which differs from one country to another and compares modern tramways with modern buses, the tram would appear to offer greater advantages as a means of public transport. It can consist of trailer-combinations and occupies the least amount of road space per passenger. It may also be possible in suburbs to run tramways on separate tracks and thereby to increase average speeds as well as reduce the number of accidents and also the volume of traffic using the main exit roads. The disadvantages of a rail-bound means of transport in town centres and of the islands which are necessary for the safety of boarding and alighting passengers would hardly weigh more heavily within a properly regulated urban transport system than those of buses which have to cross the stream of traffic twice at every stop so long as there is no special lane reserved to them. Since the replacement of trams by buses with identical capacity leads to a considerable increase in the number of vehicles the pollution of the air by exhaust gases which is particularly noticeable in the narrow streets of town centres should by no means be overlooked. On the other hand, the operation of buses in rapidly growing towns can be adapted more easily to changes in the concentration of the population than the tramways. Bus services are consequently particularly suitable for the introduction of new connections.

Taxi services are of considerable importance in connection with urban traffic planning. Similar to public transport taxi traffic also aims at the greatest possible utilization of capacity through continuity of operation without long stops or parking. It should consequently receive preferential treatment in comparison with individual transport within any procedure of granting concessions. The number of taxis admitted to the most important traffic arteries during peak hours could, however, be restricted, if public means of transport had to receive priority in order to increase the speed of operation.

A final solution of the transport problems in large towns appears only to be possible by spreading traffic along two levels. The costs of construction involved in such schemes would, however, only be justifiable if less costly measures were no longer adequate. Experience has shown that it is mainly commuters' traffic limited to certain periods which uses underground facilities. This also applies to elevated railways the construction of which in urban areas does, however, furthermore involve aesthetic considerations. The remaining day-time traffic rather prefers vehicles which are accessible without additional effort and afford a view of the surroundings. The question may therefore arise whether it might not be preferable to build underground roads to carry part of the individual motorized traffic.

We do not wish to go into any further details in this connection and only refer to the useful reports published by the International Union of Public Transport* as a result of comprehensive studies.

In conclusion we believe that a regulation of urban transport will only be justified from the point of view of the national economy and that of the transport industry if it limits individual and goods transport to the extent required in the interest of the maximum efficiency of public passenger transport.

3. The remunerativeness of urban transport enterprises

The principle of economic self-sufficiency may only be applied to urban transport enterprises to the extent to which they are offered an opportunity to economically operate their transport services. This will not be the case wherever vehicles have to force their way slowly through congested roads and where consequently additional costs arise or wherever enormous investments in rolling stock are necessary in order to cater for peak requirements, with the same vehicles remaining unused to all intents and purposes outside these peak periods. In many instances the remunerativeness is also adversely influenced by a social tariff structure involving considerable rebates for workers and school children.

In similar circumstances efforts aiming at economic self-sufficiency may have disastrous consequences. The considerable initial capital outlay involves relatively high overhead costs for depreciation and capital interest which can lead to prohibitive tariffs if they are passed on to the fares. The demand for services of urban transport undertakings is comparatively elastic—at any rate wherever passengers are carried at normal fares—because it is comparatively easy for them to change over to individual means of transport or to refrain from using public transport. If therefore rates only tend to fully cover average costs without at the same time taking into consideration the possible reaction on the part of those demanding transport services to the level of fares the danger may arise that enterprises with high fixed costs and rates which are only intended to cover costs might manoeuvre themselves out of the market.

The high share of fixed costs particularly in the case of tramways and underground railways entails a distinct digression of costs if a high degree of utilization of capacity is obtained. If, however, rates approach costs it would be precisely such enterprises which would fail to obtain a maximum utilization of capacity if the high rates lead to a gradual falling off in demand. Admittedly an increase in fares for the purpose of improving the coverage of costs may still be conducive to increased revenue even with a considerably reduced demand. The possibility exists, however, that it would simultaneously lead to an increase of average costs because of the deterioration of the utilization of capacity which has occurred. This may then in the last resort balance the increased revenue resulting from increased rates in view of the increase

*International Union of Public Transport; Congress documents 1955 and 1957.

in average costs caused by the slacking-off in demand or even exceed them. Economic self-sufficiency can therefore frequently not be achieved in the long run despite increases in rates.

Attempts to apply the principle of economic self-sufficiency may consequently in many instances lead to a vicious circle involving an increase in rates and average costs per traffic unit of urban transport undertakings which would finally not only lead to chronic deficits but in the absence of an efficient regulation of traffic also to an even worse congestion of the roads.

We believe in principle that economic self-sufficiency should nevertheless be aimed at from a commercial point of view on grounds of a rational management of urban transport undertakings. It should, however, at the same time be realized that such endeavours cannot succeed in the long run in the absence of a regulation of traffic. Within the framework of the overall urban economy remunerative management in many instances falls far short of optimization. In this instance transport is an element of the national economy as a whole and has also got to be considered in this perspective. Within the national budget the deficit of a transport enterprise may be more than balanced by higher productivity in other industries so that in reality the collectivity derives a benefit. It is consequently no more than fair for the public authorities to cover these deficits in order to prevent them exercising a pressure on the social conditions of the personnel.

It is conversely an anachronism if even today transport undertakings in many towns are obliged to pay dues for the use of roads. This conception dates back to times where such enterprises were intended to occupy some sort of position of monopoly which involved a surplus of revenue. It has been out-dated for a long time and any duties levied for the use of roads would inherently only at best be appropriate in the case of individual transport which represents a burden on the roads but not in the case of means of public transport which relieve pressure on the roads.

If it is a question of the public authorities recovering the amount of deficits of urban transport undertakings they have a number of possibilities at their disposal. In the first place possibilities exist for the saving in construction by means of an appropriate regulation of traffic. If, for instance, the stream of individual transport can be induced by means of attractive fares to park its vehicles in the suburbs and consequently relieve pressure on the town centres the loss in revenue of transport undertakings to be covered by the public authorities might still be lower than the saving in costs for improvements of roads and safety installations. We believe on the other hand that it would be justified to impose on that part of individual transport which does not avail itself of similar possibilities and thereby continues to contribute to congestion in the town centres certain special fees as a contribution to the expenditure occasioned by it, provided there are no special reasons for granting exceptional concessions.

The possibility has frequently been considered to levy a special tax on the increase in value of real estate in the urban areas produced by the existence of an efficient transport undertaking in order to cover the costs which are caused thereby. In our opinion a measure of this kind would meet with administrative difficulties because it would be necessary to investigate in this connection to what extent such increase in value would in each individual case be a consequence of the functioning of the means of public transport in question. Furthermore real estate in town centres to which preference is given in view of the position is frequently acquired by the public authorities for administrative purposes and would therefore have to be exempted from such tax.

The system of covering a deficit arising in the operation of transport by public authorities and its being passed on to the tax-payer ought to be at any rate replaced within the framework of an urban regulation of traffic. There is little point in burdening transport undertakings with charges expressed in fixed costs which no longer represent present-day conceptions on the one hand and on the other hand expect them to accept obligations in the nature of a public service on grounds of traffic regulations and social policy in order to subsequently cover the deficit which has resulted on expiry of the financial year. Urban transport is as much an essential public service as power supplies, police, or health services. In view of this consideration in the interest of the overall national economy it ought to be equipped accordingly and not required to accept any greater obligations than those which it may be expected to fulfil from the point of view of its own economic remunerativeness.

X. CIVIL AVIATION

The problems of civil aviation may only be treated in an incomplete way within the framework of a report on the question of coordination of transport. This is due to the fact that the effectiveness of measures and bodies of coordination are initially bound to be restricted to national limits, whereas civil aviation, by its very nature and particularly within Europe, extends beyond national frontiers. The problem of coordination therefore arises within the national frontiers as well as at the international level.

In numerous European countries, there is a leading national airline which is largely publicly owned. Even where there are several large companies, as for instance in Great Britain, they only represent different branches of activity of the States in the field of civil aviation.

The national airlines of Europe which were at least partly founded because of considerations of national prestige have obtained such an overwhelming importance in civil aviation that the question of their coordination with smaller private companies will in actual practice hardly arise. There is therefore a difference between conditions in Europe and in the United States where there are numerous operators in the field of civil aviation and where, consequently, the problem of coordination within that branch of the transport industry had already arisen at an early date. For the purpose of such coordination, a special body—the Civil Aeronautics Board—had been created in the United States already before the second world war.

The problem which is of primary importance in Europe is that of coordination between civil aviation and the remaining branches of the transport industry on land and sea. This problem cannot easily be divided into national and international ones because competing services as a rule extend over longer distances. The questions which have to be dealt with in this connection consequently concern national as well as European aspects and create a special problem of integration.

In the following chapters we are going to restrict ourselves to the discussion of three factors: competition between civil aviation and the branches of the inland transport industry, mainly the railways; existing distortions of competition; and, finally, regulating measures which in our opinion ought to be adopted.

1. Competition between civil aviation and inland transport

Civil aviation and the repercussions of its competition gravitate towards passenger transport, although the transport of expensive and very urgent consignments of goods by air has become more important during the last few years. Even heavy goods which must be urgently despatched may become eligible for air transport due to the saving in time which may be obtained thereby. The problem of the cost of conveyance is in similar cases admittedly only of secondary importance.

In passenger transport, too, the actual service which is offered is a largely decisive factor. Not only speed as such plays a role but also reliability and the coordination of timetables. In European civil aviation, the speed between traffic centres enables this branch of the transport industry to be in a far more advantageous competitive position compared with other means of transport. This advantage is, however, reduced by the time-wasting feeder services and long waiting times which delay the actual operation from point to point and by inadequate organization of passenger handling, and consequently increase the total travelling time. It is fairly safe to assume that in the case of distances up to approximately 180 miles (300 kms.) the superior speed of aircraft is practically offset by these delaying factors.

In the case of longer distances where the air transport is faster, problems arise in connection with fares. Comparisons between air fares of the tourist class and first-class railway fares[1] including sleepers in the year 1953 showed that in the case of distances exceeding approximately 300 miles air transport was already in a more favourable position than the railways along approximately one–fifth of the Western European routes which were examined. Almost forty per cent of the routes which were investigated showed that air transport was no more than fifteen per cent more expensive than the above-mentioned type of rail travel. If one adds auxiliary amenities such as meals served free of charge, possibilities of purchases of various items free of duty, etc., the difference which remains becomes even much less significant.[2] It is hardly exaggerated to say that air transport has now become a serious competitor of first-class rail travel over distances of approximately 280 miles. The only restrictive qualification which much be applied to these comparisons is that comfort and reliability, the latter particularly during the winter months, tend to become somewhat debatable.

The introduction of T.E.E. trains on particularly important long-distance services of the European Continent and similar measures within the different countries show that the railways do not intend to give up the hitherto profitable long-distance traffic without a fight. It is on the contrary reasonably safe to assume that attempts will be made to regain lost ground by means of reductions of fares or by making better provision for the second-class passenger and perhaps even by certain additional amenities. Should the passenger traffic of the railways suffer losses due to the competition of the airlines the danger of the railways trying to compensate their losses via the goods traffic could arise. This could however, lead to an increase of rates with consequent repercussions on the standard of living to the detriment of the community.

Faced with these long-distance projects of the railways, civil aviation

[1] In that year (1953) there were still three classes in passenger transport. The comparisons mentioned here refer to what was then the second class, which now corresponds to first class.

[2] *Cf.* S. Wheatcroft, Economics of European Air Transport, London, 1956, p. 143.

has shown a tendency to differentiate between the services by means of an ever-increasing division in classes whereby a somewhat privileged position is created for the passenger who is sufficiently willing to pay. On the other hand, a simultaneous attempt is being made to make air transport attractive to classes of the population with more modest means by the application of the principle of special rates for large quantities as applied by the railways. This is done by reducing the various amenities to a minimum. Whether the air traveller will in the long run be prepared to put up with these austerity-standards of service and amenities and to travel in sardine-fashion is, of course, another matter. The repercussions of any exaggerated tendencies in that direction may be felt particularly once flying has lost the thrill of the novelty for large groups of the population.

The only obstacle which limits the endeavours of the airline to pander to the whims of the customer to an ever-increasing extent is the increase in the cost of civil aviation. Today the level of costs per ton and/or passenger-kilometre in European scheduled air transport is approximately one and a half times higher than on comparable distances in the United States. This difference is mainly due to a considerably higher pay load and the many times higher frequency of the various services of the U.S.A. The lower remunerativeness of European services on the other hand appears last but by no means least to be due to the fact that the network of lines is too dense and that furthermore too small aircraft are used on a number of routes. The result is frequently a competition between State-owned airlines and State-owned railways whereby the deficits which they inflict on each other have to be covered from public funds.

2. Elements of the distortion of competition

The direct and indirect subsidizing of air transport by the State are the main elements of the existing distortions of competition.

Direct subsidies may be justified wherever important national airlines have to be created or reconstructed. Assuming that air transport, too, had to supply proof of its economic feasibility in comparison with other branches of the transport industry, subsidies of this kind should not be granted for any longer period than absolutely necessary. This applies particularly to grants for the carriage of mail which in the majority of cases appear to exceed by far the amount justified by the actual costs.

Indirect subsidies result in the first instance from construction and maintenance of airports. Airport charges as a rule only cover a fraction of the actual cost. This factor is of particular importance today because the impending operational use of turbo-jets is expected to entail additional vast expenditure. The distortion of competition will consequently, instead of initiating a trend in the direction of economic self-sufficiency and the concomitant covering of track costs, exert an influence in the opposite direction. The fact that certain countries now levy special airport charges on the departing passenger does not change this position in any way. We rather feel that it is wrong to arbitrarily charge passenger traffic which is part of the total airport traffic with additional costs which really ought to be included in the air fare.

Further distortions of competition of minor importance may be caused by the fact that safety installations are subject to the control of public authorities who in common with municipal authorities contribute towards the costs of these installations. Training of airline pilots also largely constitutes a charge on the State. Generally speaking, civil aviation does, however, profit from the military aspects which are connected with it. It is equally true that the costs of training personnel of other industries are also frequently entirely or partly covered by the State.

These factors prevent a fair competition with the branches of the inland transport industry, particularly the railways. Within the framework of coordination of transport the principle of economic feasibility should, however, also be realized in civil aviation. In view of the rapid development of this branch of the transport industry and the investments which thereby continuously become necessary the application of the aforementioned principle may admittedly only be considered as a long-term objective. We believe, however, that an increased share of track costs could without any difficulty be included immediately in air fares in order to arrive at a justifiable differentiation of the costs of conveyance for the benefit of inland transport.

3. Regulating measures which may be envisaged

In the majority of European countries internal air connections are of comparatively minor importance and frequently only sections of international services. A common market for national air transport only exists in exceptional cases. Internal services are as a rule only operated by the national airlines.

The importance of internal air services as feeder lines for international and inter-continental services is difficult to assess. We rather feel that considerations of prestige and wrong economic evaluations have frequently led to an over-estimation of the importance of long-distance traffic. A reduction of the excessively dense European network on the one hand and the operation of more economical types of aircraft on the remaining services would therefore appear advisable.

Conversely, the example of the United States shows that the activities of the coordinating body created by virtue of the 1938 Civil Aeronautics Act for civil aviation have, and not in the last instance, resulted in a more economic working of the airlines. This body is responsible for the approval of scheduled services. The development in Canada follows similar lines.

In Europe the creation of a body responsible for its entire territory would appear to be useful. It would have to watch the density of the network in order to safeguard economic operation and to prevent too many airlines flying along one and the same route. It would finally be the duty of such a body governed by overall considerations of transport policy, to prevent an exaggerated competition with other branches of the transport industry and the development of surplus capacities.

Attempts to introduce regulating measures in international civil aviation have up to the present been distinctly in the nature of cartels. Conferences of I.A.T.A. have as a rule restricted themselves to private rate agreements; any competition based on fares was thereby eliminated. As a result airlines have been trying to outdo each other in the field of services and amenities for the passengers which caused a forced increase of costs. Attempts to prevent such exaggerated conceptions of service by means of special agreements have only been made in exceptional cases. The relative rigidity of the level of fares has moreover been conducive to an over-estimation of the possibilities of profits and to a forced extension of capacities—a normal consequence of any cartel agreement.

Similar developments cannot be prevented by private agreements. The above-mentioned coordinating body could, however, through its influence on the tariff structure safeguard some elements of competition based on fares at least to an extent which would prevent a mutual forcing up of the costs of services and amenities.

We believe that only a coordinating body with adequate authority and responsible for a large area of Europe would be in a position to establish effective contacts with similar bodies overseas. These contacts should also be used in order to compel airlines outside Europe who fly over European territory by virtue of the "fifth freedom" and who circumvent the problem of cabotage due to the plurality of European states to contribute towards investments in ground installations. By the same means countries outside Europe could be induced to grant cabotage rights to European airlines.

XI. COASTAL NAVIGATION AND INLAND TRANSPORT

In a number of countries there is competition between coastal navigation and inland transport, particularly so far as national coastal navigation, i.e. between ports of one and the same country is concerned. Whereas the railways are in a position to offer special rates for traffic between ports and can avoid any adverse repercussions by means of their internal balancing of accounts such possibilities hardly exist in coastal navigation. The railways are, on the other hand, at any rate compelled to apply special rates in view of the competition by road haulage services which are running in parallel with railway lines. The position of coastal navigation becomes even more difficult by virtue of the fact that in certain countries the rates quoted by the railways for carriage to and from the ports involving trans-shipment are comparatively high.

The rates policy in coastal navigation is largely determined by agreements between the owners (conferences). The rates obviously have to cover costs since not only possibilities of internal balancing of accounts are lacking but also because losses of revenue are hardly likely to be covered by State subsidies.

A coordination of national coastal navigation and inland transport might be feasible as long as the former is a preserve of the national flag. This possibility should, however, not be over-rated because national coastal navigation is, as a rule, closely interwoven with international coastal navigation. Scheduled services within a given country are frequently part of international scheduled services and there is a daily change-over of tramp ships from national trades to international trades.

Wherever coastal navigation is open to foreign ships coordinating measures agreed upon at national level could only be applied within certain limits so as not to impede the participation of foreign flags and provoke counter-measures on the grounds of flag discrimination. It would consequently hardly be possible to go beyond the fixing of minimum rates.

Should agreement be reached within a group of countries on the creation of a common transport market in inland navigation considerable coordinating work could be done by means of agreements between the competent bodies of the branches of the inland transport industry on the one hand and the conferences of regular shipping lines and the bodies of tramp shipping on the other.

A similar opportunity may arise in the course of development of the European Economic Community and particularly within the framework of a Free Trade Area, always provided, however, that a basis of a common transport policy beyond the limits of inland transport can be found within these great economic areas and the appropriate transport authorities be appointed.

XII. PROBLEMS OF COMPETITION BETWEEN SEA PORTS

1. Introduction

Competition between sea ports is inherently different from that between branches of the transport industry. Whereas in competition in inland transport the demand is mainly influenced by productivity and price it is the attractiveness of the port which, apart from numerous other factors, influences the volume of traffic. Sea ports as a link between sea- and inland transport may exert a major influence on the creation of traffic centres and the direction of the main flows of traffic. It is mainly goods traffic which is involved. The importance of passenger traffic through the sea ports has been considerably reduced by the development of civil aviation.

In contrast to inland transport which could be coordinated in accordance with our explanations in the first part of our study, sea traffic is equally inaccessible to both national and international coordinating measures. The sea ports consequently occupy an intermediate position between inland transport which is capable of coordination and the elements of competition in shipping on the high seas which is essentially governed by private agreements; these elements are largely outside the scope of any coordinating influence.

In the following chapters we are going to consider the most important among the numerous factors which can exert a decisive influence on the attractiveness of a port, namely: special rates for transport to and from sea-ports, sea freights, costs of trans-shipment, port facilities, and political factors. These questions are of general importance for the policy governing sea ports. In view of the fact, however, that it is particularly the problems of competition between the North Sea ports which time and again arise within Europe, we are going to concentrate on certain aspects connected with them. Apart from wages, there is little difference between social conditions in the North Sea ports. Considerably greater inequalities within the social field would have to be taken into consideration if southern Europe ports or ports outside Europe also had to be included. It is obvious that consequently the problems of competition would also arise in different and frequently also more acute form.

2. The development of North Sea ports

A comparison based on statistics relating to the total traffic of Belgian, Netherlands, and German North Sea ports shows that the ports of the Netherlands and also, to a lesser extent, Belgian ports have increased their volume of traffic to a considerably greater extent than the German ports since the end of the war. This applies particularly to parcels- and miscellaneous goods traffic which is of major importance to the employment position in the sea ports due to its labour intensity.

DEVELOPMENT OF TRAFFIC IN BELGIAN, NETHERLANDS, AND GERMAN SEA PORTS[1]

(Period 1936-38 = 100)

(a) *Total traffic*	1952	1953	1954	1955	1956	1957
Belgian ports ...	102.3	104.0	105.0	120.6	140.6	137.0[2]
German ports ...	77.8	76.1	91.1	108.0	125.3	126.8
Netherlands ports	101.5	103.3	120.4	160.0	178.8	186.0[3]
(b) *High-value parcels and less-than-wagon-load traffic*						
Belgian ports ...	110.4	112.0	113.5	131.3	161.3	
German ports ...	90.6	81.0	100.8	119.3	142.5	
Netherlands ports	103.4	103.3	117.8	165.1	192.0	
(c) *Bulk-goods traffic*						
Belgian ports ...	92.0	94.0	94.3	107.6	114.5	
German ports ...	54.3	67.0	73.6	90.3	94.0	
Netherlands ports	94.8	103.2	130.2	141.5	131.0	

[1] The following ports of the various countries have been considered:

Belgium: Antwerp, Ghent;
Germany: Emden, Brake, Nordenham, Bremen, Bremerhaven, Hamburg;
Netherlands: Rotterdam, Schiedam, Vlaardingen, Maassluis, Dordrecht, Amsterdam.

[2] Antwerp only.

[3] Rotterdam and Amsterdam only.

In connection with the above figures attention should be paid to the fact that the total volume of traffic appears inflated by the particularly marked increase in mineral oil traffic. If this traffic were not taken into consideration the difference would be considerably reduced, as shown by the undermentioned figures relating to the years 1952 to 1955.

(d) *Total traffic without mineral oil*	1952	1953	1954	1955
Belgian ports	90.6	92.0	90.8	103.2
German ports	75.2	71.5	83.3	100.1
Netherlands ports	79.8	76.5	84.0	116.5

It is obvious that all North Sea ports have suffered considerable losses on certain pre-war services since the end of the war due to the repercussions of the political division of Europe. Whereas in the German Federal Republic mainly the port of Hamburg was affected by the partition, the Netherlands and Belgian ports have largely lost the "horseshoe traffic", i.e., the transit traffic from and to the Baltic sea ports of the now Polish and Soviet occupied areas. In addition certain structural changes in traffic have to be taken into consideration such as the strong recession in the sea-borne German coal export trade. Due to these factors the total volume of traffic available to the North Sea ports has been reduced, and consequently more serious problems of competition have to be faced.

We believe that competition between the sea ports ought to be governed by the general principle that any development ought to be avoided which could entail the redundancy of useful installations likely to remain useful in the future, otherwise there might be unemployment in ports the position of which has been artificially impaired by political developments. In application of this principle a general abolition of distinctly discriminatory practices in competition between the sea ports would appear to be in the best interests of all concerned.

3. Costs of transport to and from ports (special sea port rates)

Given otherwise equal conditions the natural geographical limits of the area in and around a sea port are influenced by the level of costs of transport in inland traffic. With differing costs of transport each branch of the transport industry has certain preferential areas for certain goods which are determined by its average costs (special preferences of traders for certain branches of the transport industry which also play a rôle so far as their preferential areas are concerned are not being referred to in this connection). The branch of the transport industry which is in a position to carry a certain merchandise or groups of merchandise at lowest average costs—in comparison with the other branches—will, all other conditions being equal, have the relatively largest preferential area.

Geographical conditions, the structure of location and density of industries also influence, apart from other factors, the level of costs of transport installations, as well as the traffic frequency of inland transport and also have a bearing on the differences between the distances to be covered to and from sea ports by one and the same branch of transport. A dense network of industries established in the hinterland of a sea port offers the advantage of relatively cheap costs of transport in traffic to and from the sea port in question, in view of the traffic frequency, whereas a mountainous and thinly-populated hinterland will cause an increase in the cost of transport to and from such ports.

It follows that the natural limits of the area served by a sea port will be determined so far as a certain merchandise or groups of merchandise are concerned, by the average costs of the cheapest branch of the transport industry which serves the port in question, all other conditions being equal, in comparison with the average costs of the branches of the transport industry serving competing sea ports. Differences in average costs which stem from distortions of the conditions of competition of the branches of the transport industry also distort the geographical limits of the areas served by the sea ports. In this respect we should like to refer to the chapter on conditions of competition in Part One of this study.

Traders do not base their decisions on comparisons between costs but between the rates quoted by the branches of the transport industry. The tariff policy may consequently cause far-reaching repercussions on the geographical extension of areas around sea ports. Wherever a branch of the transport industry is compelled by reasons of the national economy to apply rates which only serve the purpose to attract traffic to a certain port but do not cover costs, a rates policy of this type is bound to lead to a distortion of conditions of competition. It may, of course, be argued that it is precisely the railways which are frequently compelled in competition by their high overhead costs to adapt their policy rather more to market conditions than to average costs. This is particularly true in the case of competitive rates which are applied in the interest of certain limited objectives, where the railways as major enterprises are faced with a fragmented regional competition and apply these rates in order to secure for themselves their share of traffic to and from the ports. Even if such rates do not always fully cover costs they may nevertheless benefit the railways commercially if they lead to a better utilization of capacity, and at the same time additional revenue. An efficient railway will consequently always be in a position to counter-balance the natural disadvantages of a sea port by correspondingly favourable special rates. These rates are not intrinsically comparable to subsidized rates if the railway concerned were liable to suffer a loss by not applying them.

The argument against similar competitive rates is that they amount to discrimination as long as they are only granted in traffic with national sea ports. We believe, however, that in this respect similar criteria ought to be applied as in the first part of our study in respect of the geographical equality of rates. The limits within which they may be applied are reached as soon as losses would be caused to the branch of the transport industry concerned by the application of such rates. This also applies to special sea port rates. It would be contradictory to the principle of economic self-sufficiency and competition if the intrinsic interests of a branch of the transport industry would have to take second place behind the competitive interests of competing branches of the transport industry. It follows that competitive rates in sea port traffic are not of a discriminating nature if their application results in higher revenue for the branch of the transport industry concerned than if the same rates were applied to foreign sea ports.

If, for instance, a direct rail connection from the location of a given trader to sea port "A" were only possible in traffic across frontiers, i.e., involving two or more railway administrations, but on the other hand the national sea port "B" were directly accessible, the railway concerned may not be expected to apply equal rates to the foreign sea port "A" as long as its share in the freight resulting from this traffic across frontiers is smaller than the freight rate to the national sea port "B". The application of equal rates to both ports would, however, be compatible with the interests of the railway if the proportion of freight as far as the frontier in the direction of sea port "A" were equal to the total freight to port "B". The question would admittedly arise whether from a long-term point of view the repercussions of the application of equal rates to foreign ports on the national port (traffic recession) might not be far more detrimental to the interests of the railway than any possible loss of additional revenue.

4. Sea freight rates

In the majority of cases the share of sea freight in the total costs of transport which exert a major influence on the choice of a port is larger than the shares of the costs arising in connection with transport to and from ports. Sea freights consequently play an important rôle in competition between ports. Since they are entirely subject to private agreement at international level they cannot be influenced in any way by the State.

So far as sea ports are concerned a distinction must be made between tramp rates and scheduled rates. The first are formed in free competition on the international market and are consequently—at least so far as short-term charters are concerned—of a rather unstable nature, depending on peak demands at any given time. Only long-term charters in tramp shipping are comparatively stable. Freight rates in regular trade are fixed on a market which approaches the nature of a monopoly and is controlled by the "shipping conferences"; they are consequently largely stable.

The "conferences" of shipping lines between North America and Europe include the various geographically more or less separated sea ports on both ends of the lines in so-called "ranges". A range is a group of ports which may comprise so-called base ports and other ports. Between the base ports at both ends of the line concerned there is equality of freight rates, although there are frequently very considerable differences between the distances separating the possible trades within a range.

A number of "conferences" of the North America/Europe trade do not consider the German sea ports within the Antwerp-Hamburg range as base ports and consequently calculate higher freight rates for German sea ports than for trade with the Benelux ports. Supplements charged on certain goods—the so-called range supplements—appear to represent, as it were, counter-measures of the "conferences" against the sea port rates policy of the German Federal Railways. The "conferences" seem to feel that this rates policy operates to their disadvantage. A solution is only likely to be possible by mutual agreement. The argument used by the "conferences" to justify these range supplements, namely the longer distances to German sea ports, is at any rate hardly valid in view of the fact that in other cases considerably greater differences in distances are tolerated without range supplements. The question will admittedy arise whether in principle any equality of rates applied in conjunction with very considerable differences in distances is compatible at all with the normal principles underlying the fixing of rates.

5. Costs arising in ports

There are on the one hand pilot fees which shipping companies have to pay for pilots' services and port and harbour dues of varying kinds on the other.

Harbour dues include tonnage and quay dues. The method of calculating these charges differs from one port to another. Tonnage dues which are levied to defray the costs of using the harbour basin are based either on the tonnage of the ship or the quantity of commodities loaded or unloaded. Frequently the type of commodity is also taken into account. In the majority of ports quay dues are merely fees for the anchorage of the ship. In certain ports it also includes a dockage charge for the use of the quay in connection with loading and unloading operations.

In accordance with the varying methods of calculating fees and charges and the differences between wages and prices in the various countries these fees differ in each case. In addition, the levels of costs of the various ports differ considerably according to their structure. Roads leading to and from the ports may be longer or shorter, ports situated behind gates have higher costs than open ports, more dredging may have to be done in one harbour than in another and so on. Differences may also arise by the fact that a greater or smaller volume of traffic in a given port may cause a different degree of utilization of the capacity of the installations.

In our opinion the principle of economic self-sufficiency also ought to be applied to sea ports. Since sea ports are points of transit from one system of tracks to another harbour dues should be considered as some sort of coverage of track costs. Their amount would have to be fixed also with regard to market conditions. Such track costs could either be charged entirely to the shipping industry or spread over all branches of the transport industry and other cost elements. In the majority of ports all these methods are applied simultaneously. Frequently both ocean and inland navigation also cover part of the costs of harbour basins, whilst the railways are charged fees for using the quays. In addition, part of the costs of roads leading to the ports is in many instances covered by public authorities.

Handling costs consist of the costs of the installations and the social costs of dock labour. Whereas in German ports these costs are passed on to the goods which are handled, they are, in Antwerp, largely a charge on the shipping companies and in Dutch ports they are as a rule divided between the shipping companies on the one hand and shippers and consignees on the other. These differences in the methods of charging handling costs also exert an influence on the conditions of competition between sea ports. In addition, the widely divergent conditions of ownership of harbour installations have to be taken into account. In Dutch ports private loading and unloading facilities exist side by side with those under public ownership; in actual practice, however, the latter are exclusively hired out to private firms. In German ports, on the other hand, loading and unloading operations are carried out by public warehouse companies, whilst in Belgian ports these operations are partly carried out by private firms and partly by municipal undertakings.

Differences in technical installations, in the position of harbour basins and quays, in labour productivity, and finally in wages and prices of the various countries also contribute to very considerable differences in handling costs. On the whole, however, their amount does not matter a great deal because shippers and consignees choose ports according to the total costs which arise and of which handling costs are no more than a relatively unimportant fraction. They may, however, play a major rôle in the case of parity of sea freights and costs of transport to and from ports.

6. The waiving of charges or refunds of costs

There is a particular type of distortion of competition which has frequently been alleged to exist but which at any rate is difficult to prove, namely, the practice of refunding costs of transport to and from ports and other charges more or less under the counter and/or waiving them completely. Similar practices would constitute a contravention of the obligations implied in the "shipping conferences" and would artificially influence the competitive position of ports.

Both the transport industry and the shipping companies deny, however, that such refunds or waiving of fees or dues really happen. It is also perfectly feasible that certain forms of refunding part of the costs are applied as a matter of normal commercial usage. Certain shipping companies are, for instance, shareholders of enterprises in inland transport, as, for instance, in Rhine navigation whereby numerous possibilities of mutual agreements of this type can obviously arise.

7. The importance of facilities

Last but by no means least the attraction of a sea port in relation to the different traffic connections is influenced by the existing facilities. In this connection five main factors have to be considered. In the first place there are the natural advantages of a port resulting from its position in relation to the sea and the hinterland. Secondly, the existence of efficient loading and unloading installations as well as the cooperation of railways, road haulage, and inland navigation within the sea port. Particularly the latter implies in many instances by itself alone considerable cost reductions because direct trans-shipment from seagoing ships to inland navigation craft saves quay and storage dues. A further important factor consists in the existence of the largest possible variety of shipping possibilities in conjunction with world-wide shipping connections and a structurally balanced seaward ship- and goods traffic, coupled with possibilities of transit traffic to other countries. Commerce and trade of the sea port towns themselves and particularly those with millions of inhabitants is also of considerable importance and affords the port in question, especially if high-value goods are handled, a certain independence of inland transport costs, and frequently also a large share in the traffic of the hinterland of competing ports. Finally, the auxiliary trades in the widest sense of the word also count among these facilities, in the first instance major forwarding enterprises which almost exclusively control groupage- and container traffic as well as the existence of other auxiliary installations, e.g. repair docks.

8. Politically influenced factors of competition

Competition between sea ports is largely also politically influenced in view of the fact that the competitive position of national ports is frequently artificially improved by government measures. The repercussions of such measures amount to discrimination against foreign competing ports.

Measures which involve preferential treatment of national ports may be adopted within the framework of commercial policy, in the field of taxation or of transport policy. Trade may be encouraged to use national ports by means of reducing the costs of transport to and from them or by restricting the free choice of the shippers.

Regulating measures in the field of commercial policy consist in clauses in trade agreements which stipulate that imports or exports have to be effected f.o.b. (free on board) or c.i.f. (cost, insurance, freight). Apart from favouring national shipping companies this method in actual practice also involves advantages for national ports. The German export trade of coal and coke to Scandinavia, for instance, is governed by a clause according to which sales may only be effected f.o.b. in German ports.

Measures in the field of taxation imply reduced taxation on commodities which are imported or exported via national ports. Import and export trade as well as goods handling via the German sea ports, for instance, are exempt from turnover-tax (sea port privilege). This exemption does not apply to traffic via the ports on the Rhine and traffic to the Benelux ports consequently becomes more expensive.

In France a "surtaxe d'entrepôt" is levied on commodities which are imported via foreign sea ports. In inland navigation from Antwerp to Strasbourg as well as in regard to rail and road transport across other frontiers into France, Antwerp is placed on an equal footing with French sea ports. So far as Rotterdam is concerned this equalization only applies to Strasbourg-bound goods traffic on the Rhine.

Measures in the field of transport policy may also consist in the allocation of traffic to national and foreign ports by means of a quota system. This applies, for instance, to the sugar and corn trade to

German ports which is subject to an ordinance of the Federal Government according to which the bulk of such traffic must be directed via German ports.

Other measures in the field of transport policy exert an influence on rates. Distances to and from national ports, for instance, are frequently artificially reduced in order to reduce freight rates. Although such rates may be of a competitive nature from the point of view of the branch of the transport industry in question, they represent subsidized rates so far as the favoured port is concerned. In this connection the former German "zero-level" system and the system of "port equalization" also ought to be mentioned. Whereas the "zero level" system, as applied to railway rates still took into account the actual distances between German ports and hinterland, this is no longer the case with "port equalization." According to that system freight rates to and from all national ports should be fixed at the same level disregarding actual distances in order to artificially eliminate the natural geographical advantages of certain ports.

Railway rates to inland ports in traffic to foreign ports which involves trans-shipment are also frequently considerably higher than rates to national sea ports; or canal dues and towage fees in traffic to and from national ports are fixed lower than on waterways to foreign ports. The railways often also calculate different demurrages in order to afford preferential treatment to traffic in national ports; and finally, a very important possibility of influencing the competitive position in the field of transport policy exists wherever deficits of port administrations are covered from public funds.

PART THREE

Transport problems arising in the course of the economic integration of Europe

XIII. TEXTS OF THE TITLE OF THE TREATY CONCERNING TRANSPORT, WITH COMMENTS AND OTHER RELEVANT TEXTS

A. General Remarks

In accordance with the Treaty, there shall be created a European Community and a Common Market. One of the primary repercussions of this development is the introduction of a common policy in the field of the transport industry.

Article 61 of the Treaty stipulates that the free exchange of transport services shall be governed by the provisions of the Title relating to transport (Part II, Title IV).

Under this heading it is worth stressing in the first instance that the Treaty provides *inter alia* for the following bodies which are to be entrusted with carrying out the different tasks:

(a) A Consultative Committee shall be attached to the European Commission. This Committee shall consist of experts who shall be nominated by the governments of Member States. The Committee shall be granted a hearing as may become necessary.

(b) A special Transport Section shall be attached to the Economic and Social Committee. This special section may not be consulted independently of the Economic and Social Committee. In view of the fact that in accordance with Article 198 the Economic and Social Committee *must* be consulted in a number of questions, whereas the Transport Committee mentioned under (a) *may* be granted a hearing at the discretion of the Commission, there is a possibility of the Transport Section in actual practice obtaining more extensive possibilities to make its influence felt with the Economic and Social Committee than the Expert Committee.

We must point out that the Common Market is to be realized gradually in the course of a period of transition of 12 years. This period may be extended to a maximum of 15 years. The period of transition is divided into three parts.

The first sub-period lasts four years but may be extended to six years. If it is found that the main objectives provided by the Treaty for the first stage have been largely realized and the relevant obligations fulfilled, the transition to the second sub-period will take place. This sub-period as well as the third shall last four years. Both sub-periods may only be extended or reduced by decision of the Council. The decision shall be based on a proposal of the Commission and shall be reached unanimously.

Finally we wish to point out that two problems which are of considerable importance for the transport industry are not dealt with under the Title relating to transport, *viz.* the right of establishment to which Art. 52—58 inclusive are devoted and the cartels which are regulated by the Rules on enterprises, Art. 85—90 inclusive.

B. Stipulations concerning Transport

ART. 74

The objectives of this Treaty shall, with regard to the subject covered by this Title, be pursued by the Member States within the framework of a common transport policy.

In this Article the "objectives of the Treaty" are mentioned.

The opinion has been expressed that all provisions of the Treaty should also be applicable to the transport industry. Consequently, not only Title IV would be of decisive importance. It would mean that the objective of the Treaty in accordance with Article 2, 3, 7 and 61 should be the creation of a Common Transport Market which would involve a free exchange of transport services.

According to another opinion the creation of a Common Market is, however, only intended to be the means for the realization of those objectives which result from Article 2. Consequently, the activity of the Economic Community in the field of transport, so far as the creation of a Common Market is concerned, would be determined exclusively by Title IV, and the only result in this respect would be the introduction of a Common transport policy which in turn would have to be determined again by the Council, in accordance with Article 75.

ART. 75

1. *With a view to implementing Article 74 and taking due account of the special aspects of transport, the Council, acting on a proposal of the Commission and after the Economic and Social Committee and the Assembly have been consulted, shall, until the end of the second stage by means of a unanimous vote and subsequently by means of a qualified majority vote, lay down:*

(a) *common rules applicable to international transport effected from or to the territory of a Member State or crossing the territory of one or more Member States;*

(b) *conditions for the admission of non-resident carriers to national transport services within a Member State; and*

(c) *any other appropriate provisions.*

2. *The provisions referred to under (a) and (b) of the preceding paragraph shall be laid down in the course of the transitional period.*

3. *Notwithstanding the procedure provided for in paragraph 1, provisions which relate to the principles governing transport and the application of which might seriously affect the standard of living and the level of employment in certain regions and also the utilisation of transport equipment, shall, due account being taken of the need for adaptation to economic developments resulting from the establishment of the Common Market, be laid down by the Council acting by means of a unanimous vote.*

According to one opinion paragraph 1(a) is only intended to deal with the participation in traffic across frontiers on a basis of equality. According to the other opinion, these provisions can refer to everything which is of importance for the traffic across the frontiers and/or the transit traffic of the participating States. This would, in accordance with the same opinion, also apply particularly to the introduction of direct railway tariffs.

Similar difficulties result in connection with 1(b). In accordance with one opinion, this paragraph is again only intended to refer to the participation of foreign carriers in the inland transport of the different countries on a basis of equality. According to the other opinion, the Council is completely at liberty to fix the conditions of such participation.

No difficulties of interpretation arise in connection with point 1(c). This paragraph contains the unequivocal provision that the Council may in actual practice take all useful measures, consequently also those which could still appear debatable under the other two points.

Different opinions have also been expressed as to whether the provisions of items 1(a)—1(c) are merely an enumeration or whether the sequence also implied a priority.

So far as paragraph 3 is concerned, there appears to be a certain divergence between the German text on the one hand and the French and Dutch texts on the other. Both of the last-named texts state as an indispensable condition that the Council shall issue unanimously agreed regulations at the expiry of the second stage if their application was likely to seriously affect the standard of living *and* the employment position in certain areas. According to the German text one might feel inclined to assume that *one* of these two conditions in itself would suffice.

ART. 76

Until the provisions referred to in Article 75, paragraph 1, are enacted and unless the Council gives its unanimous consent, no Member State shall apply the various provisions governing this subject at the date of the entry into force of this Treaty in such a way as to make them less favourable, in their direct or indirect effect, for carriers of other Member States by comparison with its own national carriers.

Whether this "freeze" is also intended to apply to tariff measures is by no means certain.

ART. 77

Aids which meet the needs of transport coordination or which constitute reimbursement for certain obligations inherent in the concept of a public utility shall be deemed to be compatible with this Treaty.

It may be concluded from this Article that subsidies will be admissible in so far as they are intended to further the equalization of the conditions of competition.

Art. 78

Any measure in the sphere of transport rates and conditions adopted within the framework of this Treaty, shall take due account of the economic situation of carriers.

This clearly implies that the process of integration should not be carried out at the expense of the carriers.

Art. 79

1. *Any discrimination which consists in the application by a carrier, in respect of the same goods conveyed in the same circumstances, of transport rates and conditions which differ on the ground of the country of origin or destination of the goods carried, shall be abolished in the traffic within the Community not later than at the end of the second stage.*

2. *Paragraph 1 shall not exclude the adoption of other measures by the Council in application of Article 75, paragraph 1.*

3. *The Council, acting by means of a qualified majority vote on a proposal of the Commission and after the Economic and Social Committee has been consulted, shall, within a period of two years after the date of the entry into force of this Treaty, lay down rules for the implementation of the provisions of paragraph 1.*

The Council may, in particular, enact the provisions necessary to enable the institutions of the Community to ensure that the rule stated in paragraph 1 is observed and that all the advantages accruing from it are enjoyed by users.

4. *The Commission shall, on its own initiative or at the request of a Member State, examine the cases of discrimination referred to in paragraph 1 and shall, after consulting any Member State interested, take the necessary decisions within the framework of the rules laid down in accordance with the provisions of paragraph 3.*

In connection with paragraph 1 one might ask what precisely is meant by "in the same circumstances." Another question which arises in this connection is whether traffic in various directions along one and the same line ought to be considered as "the same circumstances." The difficulties which arise in connection with paragraphs 2 to 4 partly originate in paragraph 1. According to one opinion, paragraph 1 contains an exclusive definition of the conception of discrimination and stipulates at the same time that all discrimination shall be eliminated before the conclusion of the second stage. This would imply that the only discriminations which may exist are those enumerated in paragraph 1.

Others believe that paragraph 1 does not contain any clear definition of the conception of discrimination at all and furthermore only refers to the type of discriminations which should be eliminated on termination of the second stage. This interpretation would imply that the Council could issue all useful regulations in accordance with Article 75 paragraph 1(c), also with reference to those discriminations which are not referred to at all in paragraph 1 of Article 79.

ART. 80

1. *The application imposed by a Member State, in respect of transport effected within the Community, of rates and conditions involving any element of support or protection in the interest of one or more particular enterprises or industries shall be prohibited as from the beginning of the second stage, unless authorised by the Commission.*

2. *The Commission shall, on its own initiative or at the request of a Member State, examine the rates and conditions referred to in paragraph 1, taking particular account, on the one hand, of the requirements of a suitable regional economic policy, of the needs of under-developed regions and the problems of regions seriously affected by political circumstances and, on the other hand, of the effects of such rates and conditions on competition between the different forms of transport.*

After consulting any interested Member State, the Commission shall take the necessary decisions.

3. *The prohibition referred to in paragraph 1 shall not apply to competitive tariffs.*

This Article is characterized particularly by its vague wording and its ambiguous nature. The purpose underlying this formulation appears to be the intention to largely maintain the existing position. In accordance with paragraph 1 all subsidies or protectionist measures for the benefit of certain undertakings or branches of the industry should be prohibited with effect from the beginning of the second stage.

Paragraph 3 however stipulates that competitive rates are not affected at all by this prohibition. In view of the fact that there is no detailed definition of the conception of competitive rates one might be entitled to ask exactly what tariffs are referred to in this connection.

Is it, for instance, the intention to consider exceptional rates of sea ports as competitive rates? The question which arises here is one of principle, namely whether the definitions introduced for instance by the European Community for Coal and Steel, for the conception of competitive rates, should also be applicable in this instance.

Opinions are divided on this point.

Furthermore, the Commission should be entitled to permit, after appropriate investigation, protective rates and special rates for support of industries in so far as they serve requirements of the policy governing the location of industries of under-developed areas and areas which are politically in a disadvantageous position. This refers to those tariffs which are applicable to areas without industries or with a low density of population, such as in France and Italy, as well as for areas in Germany situated near the frontiers of the Zones, and finally for the Italian/Yugoslav frontier area.

ART. 81

Charges or dues collected by a carrier, in addition to the transport rates, for the crossing of frontiers, shall not exceed a reasonable level, due account being taken of real costs actually incurred by such crossing.

Member States shall endeavour to reduce these costs progressively.

The Commission may make recommendations to Member States with a view to the application of this Article.

It is not clear whether this text also refers to the handling charges of the railways which are calculated independently from the actual freight rates.

ART. 82

The provisions of this Chapter shall not be an obstacle to the measures taken in the Federal Republic of Germany, to the extent that such measures may be necessary to compensate for the economic disadvantages caused by the division of Germany to the economy of certain regions of the Federal Republic which are affected by that division.

This so-called German clause again refers to the complex of areas near the Zone frontiers, the favoured position of which ought to be safeguarded already by virtue of Article 80, paragraph 2. It is believed that this Article 82 merely implies that similar protective measures do not require the official decision of the Commission.

Others hold that such a decision will in any case be necessary and indispensable.

ART. 83

A Committee with consultative status, composed of experts appointed by the Governments of Member States, shall be established and attached to the Commission. The latter shall, whenever it deems it desirable, consult this Committee on transport questions, without prejudice to the competence of the transport section of the Economic and Social Committee.

The Experts Committee which is attached to the Commission may only express an opinion if it is consulted by the Commission. Its powers would therefore be less comprehensive than those of the Committee of Transport Experts of the European Community for Coal and Steel (E.C.C.S.).

We should like to stress in this connection that the experts should on no account be selected exclusively from the ranks of the employers' organizations, as in the case of the European Community for Coal and

Steel. The workers' organizations are also in a position to make available expert advice.

ART. 84

1. *The provisions of this Title shall apply to transport by rail, road and inland waterway.*

2. *The Council, acting by means of a unanimous vote, may decide whether, to what extent and by what procedure appropriate provisions might be adopted for sea and air transport.*

The appropriate measures in the field of navigation and civil aviation may only be adopted by a unanimous decision of the Council, whereby the type of measures and the relevant procedure would have to be determined. The inclusion of these latter industries is likely to become important fairly soon, particularly with regard to coastal navigation and the problem of cabotage in civil aviation.

C. Further Stipulations of Interest to Transport Workers

Part 2: Title III

Chapter 1: WORKERS

ART. 48

1. *The free movement of workers shall be ensured within the Community not later than at the date of the expiry of the transitional period.*

2. *This shall involve the abolition of any discrimination based on nationality between workers of the Member States as regards employment, remuneration and other working conditions.*

3. *It shall include the right, subject to limitations justified by reasons of public order, public safety and public health:*

(*a*) *to accept offers of employment actually made;*

(*b*) *to move about freely for this purpose within the territory of Member States;*

(*c*) *to stay in any Member State in order to carry on an employment in conformity with the legislative and administrative provisions governing the employment of the workers of that State; and*

(*d*) *to live, on conditions which shall be the subject of implementing regulations to be laid down by the Commission, in the territory of a Member State after having been employed there.*

4. *The provisions of this Article shall not apply to employment in the public administration.*

ART. 49

Upon the entry into force of this Treaty, the Council, acting on a proposal of the Commission and after the Economic and Social

Committee has been consulted, shall, by means of directives or regulations, lay down the measures necessary to effect progressively the free movement of workers, as defined in the preceding Article, in particular:

(*a*) *by ensuring close collaboration between national labour administrations;*

(*b*) *by progressively abolishing according to a plan any such administrative procedures and practices and also any such time-limits in respect of eligibility for available employment as are applied as a result either of municipal law or of agreements previously concluded between Member States and the maintenance of which would be an obstacle to the freeing of the movement of workers;*

(*c*) *by progressively abolishing according to a plan all such time-limits and other restrictions provided for either under municipal law or under agreements previously concluded between Member States as impose on workers of other Member States conditions for the free choice of employment different from these imposed on workers of the State concerned; and*

(*d*) *by setting up appropriate machinery for connecting offers of employment and requests for employment, with a view to equilibrating them in such a way as to avoid serious threats to the standard of living and employment in the various regions and industries.*

ART. 50

Member States shall, under a common programme, encourage the exchange of young workers.

ART. 51

The Council, acting by means of a unanimous vote on a proposal of the Commission, shall, in the field of social security, adopt the measures necessary to effect the free movement of workers, in particular, by introducing a system which permits an assurance to be given to migrant workers and their beneficiaries:

(*a*) *that, for the purposes of qualifying for and retaining the right to benefits and of the calculation of these benefits, all periods taken into consideration by the respective municipal law of the countries concerned, shall be added together, and*

(*b*) *that these benefits will be paid to persons resident in the territories of Member States.*

Chapter 2: THE RIGHT OF ESTABLISHMENT

ART. 52

Within the framework of the provisions set out below, restrictions on the freedom of establishment of nationals of a Member State in the territory of another Member State shall be progressively abolished in the course of the transitional period Such progressive abolition shall also extend to restrictions on the setting up of agencies, branches or subsidiaries by nationals of any Member State established in the territory of any Member State.

Freedom of establishment shall include the right to engage in and carry on non-wage-earning activities, and also to set up and manage enterprises and, in particular, companies within the meaning of Article 58, second paragraph. under the conditions laid down by the law of the country of establishment for its own nationals. subject to the provisions of the Chapter relating to capital.

ART. 53

Member States shall not, subject to the provisions of this Treaty, introduce any new restrictions on the establishment in their territories of nationals of other Member States.

ART. 54

1. *Before the expiry of the first stage, the Council, acting by means of a unanimous vote on a proposal of the Commission and after the Economic and Social Committee and the Assembly have been consulted, shall lay down a general programme for the abolition of restrictions existing within the Community on freedom of establishment. The Commission shall submit such proposal to the Council in the course of the first two years of the first stage.*

The programme shall, in respect of each category of activities, fix the general conditions for achieving freedom of establishment and, in particular, the stages by which it shall be attained.

2. *In order to implement the general programme or, if no such programme exists, to complete one stage towards the achievement of freedom of establishment for a specific activity, the Council, on a proposal of the Commission and after the Economic and Social Committee and the Assembly have been consulted, shall, until the end of the first stage by means of a unanimous vote and subsequently by means of a qualified majority vote, act by issuing directives.*

3. *The Council and the Commission shall exercise the functions entrusted to them by the above provisions, in particular:*

(a) by according, as a general rule, priority treatment to activities in regard to which freedom of establishment constitutes a specially valuable contribution to the development of production and trade;

(b) by ensuring close collaboration between the competent national authorities with a view to ascertaining the special situation within the Community of the various activities concerned;

(c) by abolishing any such administrative procedures and practice whether resulting from municipal law or from agreements previously concluded between Member States as would, if maintained, be an obstacle to freedom of establishment;

(d) by ensuring that wage-earning workers of one Member State employed in the territory of another Member State may remain in that territory for the purpose of undertaking a non-wage-earning activity there, provided that they satisfy the conditions which they would be required to satisfy if they came to that State at the time when they wish to engage in such activity;

(e) by enabling a national of one Member State to acquire and exploit real property situated in the territory of another Member State, to the extent that no infringement of the principles laid down in Article 39, paragraph 2 is thereby caused;

(f) by applying the progressive abolition of restrictions on freedom of establishment, in each branch of activity under consideration, both in respect of the conditions for setting up agencies, branches or subsidiaries in the territory of a Member State and in respect of the conditions governing the entry of personnel of the main establishment into the managerial or supervisory organs of such agencies, branches and subsidiaries;

(g) by coordinating, to the extent that is necessary and with a view to making them equivalent, the guarantees demanded in Member States from companies within the meaning of Article 58, second paragraph, for the purpose of protecting the interests both of the members of such companies and of third parties; and

(h) by satisfying themselves that conditions of establishment shall not be impaired by any aids granted by Member States.

ART. 55

Activities which in any State include, even incidentally, the exercise of public authority shall, in so far as that State is concerned, be excluded from the application of the provisions of this Chapter.

The Council, acting by means of a qualified majority vote on a proposal of the Commission, may exclude certain activities from the application of the provisions of this Chapter.

ART. 56

1. *The provisions of this Chapter and the measures taken in pursuance thereof shall not prejudice the applicability of legislative and administrative provisions which lay down special treatment for*

foreign nationals and which are justified by reasons of public order, public safety and public health.

2. *Before the expiry of the transitional period, the Council, acting by means of a unanimous vote on a proposal of the Commission and after the Assembly has been consulted, shall issue directives for the co-ordination of the above-mentioned legislative and administrative provisions. After the end of the second stage, however, the Council, acting by means of a qualified majority vote on a proposal of the Commission, shall issue directives for coordinating such provisions as, in each Member State, fall within the administrative field.*

ART. 57

1. *In order to facilitate the engagement in and exercise of non-wage-earning activities, the Council, on a proposal of the Commission and after the Assembly has been consulted, shall, in the course of the first stage by means of a unanimous vote and subsequently by means of a qualified majority vote, act by issuing directives regarding mutual recognition of diplomas, certificates and other qualifications.*

2. *For the same purpose, the Council, acting on a proposal of the Commission and after the Assembly has been consulted, shall, before the expiry of the transitional period, issue directives regarding the coordination of legislative and administrative provisions of Member States concerning the engagement in and exercise of non-wage-earning activities. A unanimous vote shall be required on matters which, in at least one Member State, are subject to legislative provisions, and on measures concerning the protection of savings, in particular the allotment of credit and the banking profession, and concerning the conditions governing the exercise in the various Member States of the medical, para-medical and pharmaceutical professions. In all other cases, the Council shall act in the course of the first stage by means of a unanimous vote and subsequently by means of a qualified majority vote.*

3. *In the case of the medical, para-medical and pharmaceutical professions, the progressive removal of restrictions shall be subject to the co-ordination of conditions for their exercise in the various Member States.*

ART. 58

Companies constituted in accordance with the law of a Member State and having their registered office, central management or main establishment within the Community shall, for the purpose of applying the provisions of this Chapter, be assimilated to natural persons being nationals of Member States.

The term "companies" shall mean companies under civil or commercial law including cooperative companies and other legal persons under public or private law, with the exception of non-profit-making companies.

Chapter 1, Section 1: RULES APPLYING TO ENTERPRISES

ART. 85

1. *The following shall be deemed to be incompatible with the Common Market and shall hereby be prohibited: any agreements between enterprises, any decisions by associations of enterprises and any concerted practices which are likely to affect trade between the Member States and which have as their object or result the prevention, restriction or distortion of competition within the Common Market, in particular those consisting in:*

(a) *the direct or indirect fixing of purchase or selling prices or of any other trading conditions;*

(b) *the limitation or control of production, markets, technical development or investment;*

(c) *market-sharing or the sharing of sources of supply;*

(d) *the application to parties to transactions of unequal terms in respect of equivalent supplies, thereby placing them at a competitive disadvantage; or*

(e) *the subjecting of the conclusion of a contract to the acceptance by a party of additional supplies which, either by their nature or according to commercial usage, have no connection with the subject of such contract.*

2. *Any agreements or decisions prohibited pursuant to this Article shall be null and void.*

3. *Nevertheless, the provisions of paragraph 1 may be declared inapplicable in the case of:*

—*any agreements or classes of agreements between enterprises,*
—*any decisions or classes of decisions by associations of enterprises, and*
—*any concerted practices or classes of concerted practices which contribute to the improvement of the production or distribution of goods or to the promotion of technical or economic progress while reserving to users an equitable share in the profit resulting therefrom, and which:*

(a) *neither impose on the enterprises concerned any restrictions not indispensable to the attainment of the above objectives;*

(b) *nor enable such enterprises to eliminate competition in respect of a substantial proportion of the goods concerned.*

ART. 86

To the extent to which trade between any Member States may be affected thereby, action by one or more enterprises to take improper advantage of a dominant position within the Common Market or within a substantial part of it shall be deemed to be incompatible with the Common Market and shall hereby be prohibited.

Such improper practices may, in particular, consist in:

(a) the direct or indirect imposition of any inequitable purchase or selling prices or of any other inequitable trading conditions;

(b) the limitation of production, markets or technical development to the prejudice of consumers;

(c) the application to parties to transactions of unequal terms in respect of equivalent supplies, thereby placing them at a competitive disadvantage; or

(d) the subjecting of the conclusion of a contract to the acceptance, by a party, of additional supplies which, either by their nature or according to commercial usage, have no connection with the subject of such contract.

ART. 87

1. *Within a period of three years after the date of the entry into force of this Treaty, the Council, acting by means of a unanimous vote on a proposal of the Commission and after the Assembly has been consulted, shall lay down any appropriate regulations or directives with a view to the application of the principles set out in Articles 85 and 86.*

If such provisions have not been adopted within the above-mentioned time-limit, they shall be laid down by the Council acting by means of a qualified majority vote on a proposal of the Commission and after the Assembly has been consulted.

2. *The provisions referred to in paragraph 1 shall be designed, in particular:*

(a) to ensure observance, by the institution of fines or penalties, of the prohibitions referred to in Article 85, paragraph 1, and in Article 86;

(b) to determine the particulars of the application of Article 85 paragraph 3, taking due account of the need, on the one hand, of ensuring effective supervision and, on the other hand, of simplifying administrative control to the greatest possible extent;

(c) to specify, where necessary, the scope of application in the various economic sectors of the provisions contained in Articles 85 and 86;

(d) to define the respective responsibilities of the Commission and of the Court of Justice in the application of the provisions referred to in this paragraph; and

(e) *to define the relations between, on the one hand, municipal law and, on the other hand, the provisions contained in this Section or adopted in application of this Article.*

ART. 88

Until the date of the entry into force of the provisions adopted in application of Article 87, the authorities of Member States shall, in accordance with their respective municipal law and with the provisions of Article 85, particularly paragraph 3, and of Article 86, rule upon the admissibility of any understanding and upon any improper advantage taken of a dominant position in the Common Market.

ART. 89

1. *Without prejudice to the provisions of Article 88, the Commission shall, upon taking up its duties, ensure the application of the principles laid down in Articles 85 and 86. It shall, at the request of a Member State, or ex officio, investigate, in conjunction with the competent authorities of the Member States which shall lend it their assistance, any alleged infringement of the above-mentioned principles. If it finds that such infringement has taken place, it shall propose appropriate means for bringing it to an end.*

2. *If such infringement continues, the Commission shall, by means of a reasoned decision, confirm the existence of such infringement of the principles. The Commission may publish its decision and may authorise Member States to take the necessary measures, of which it shall determine the conditions and particulars, to remedy the situation.*

ART. 90

1. *Member States shall, in respect of public enterprises and enterprises to which they grant special or exclusive rights, neither enact nor maintain in force any measure contrary to the rules contained in this Treaty, in particular, to those rules provided for in Article 7 and in Articles 85 to 94 inclusive.*

2. *Any enterprise charged with the management of services of general economic interest or having the character of a fiscal monopoly shall be subject to the rules contained in this Treaty, in particular to those governing competition, to the extent that the application of such rules does not obstruct the de jure or de facto fulfilment of the specific tasks entrusted to such enterprise. The development of trade may not be affected to such a degree as would be contrary to the interests of the Community.*

3. *The Commission shall ensure the application of the provisions of this Article and shall, where necessary, issue appropriate directives or decisions to Member States.*

Chapter 2: FISCAL PROVISIONS

ART. 95

A Member State shall not impose, directly or indirectly, on the products of other Member States any internal charges of any kind in excess of those applied directly or indirectly to like domestic products.

Furthermore, a Member State shall not impose on the products of other Member States any internal charges of such a nature as to afford indirect protection to other productions.

Member States shall, not later than at the beginning of the second stage, abolish or amend any provisions existing at the date of the entry into force of this Treaty which are contrary to the above rules.

ART. 96

Products exported to the territory of any Member State may not benefit from any drawback of internal charges in excess of those charges imposed directly or indirectly on them.

ART. 97

Any Member States which levy a turnover tax calculated by a cumulative multi-stage system may, in the case of internal charges imposed by them on imported products or of drawbacks granted by them on exported products, establish average rates for specific products or groups of products, provided that such States do not infringe the principles laid down in Articles 95 and 96.

Where the average rates established by a Member State do not conform with the above-mentioned principles, the Commission shall issue to the State concerned appropriate directives or decisions.

ART. 98

With regard to charges other than turnover taxes, excise duties and other forms of indirect taxation, exemptions and drawbacks in respect of exports to other Member States may not be effected and compensatory charges in respect of imports coming from Member States may not be imposed, save to the extent that the measures contemplated have been previously approved for a limited period by the Council acting by means of a qualified majority vote on a proposal of the Commission.

ART. 99

The Commission shall consider in what way the law of the various Member States concerning turnover taxes, excise duties and other forms of indirect taxation, including compensatory measures applying to exchanges between Member States, can be harmonised in the interest of the Common Market.

The Commission shall submit proposals to the Council which shall act by means of a unanimous vote, without prejudice to the provisions of Articles 100 and 101.

Chapter 3: *APPROXIMATION OF LAWS*

ART. 100

The Council, acting by means of a unanimous vote on a proposal of the Commission, shall issue directives for the approximation of such legislative and administrative provisions of the Member States as have a direct incidence on the establishment or functioning of the Common Market.

The Assembly and the Economic and Social Committee shall be consulted concerning any directives whose implementation in one or more of the Member States would involve amendment of legislative provisions.

ART. 101

Where the Commission finds that a disparity existing between the legislative or administrative provisions of the Member States distorts the conditions of competition in the Common Market and thereby causes a state of affairs which must be eliminated, it shall enter into consultation with the interested Member States.

If such consultation does not result in an agreement which eliminates the particular distortion, the Council, acting during the first stage by means of a unanimous vote and subsequently by means of a qualified majority vote on a proposal of the Commission, shall issue the directives necessary for this purpose. The Commission and the Council may take any other appropriate measures as provided for in this Treaty.

ART. 102.

1. *Where there is reason to fear that the enactment or amendment of a legislative or administrative provision will cause a distortion within the meaning of the preceding Article, the Member State desiring to proceed therewith shall consult the Commission. After consulting the Member States, the Commission shall recommend to the States concerned such measures as may be appropriate to avoid the particular distortion.*

2. *If the State desiring to enact or amend its own provisions does not comply with the recommendation made to it by the Commission, other Member States may not be requested, in application of Article 101 to amend their own provisions in order to eliminate such distortion. If the Member State which has ignored the Commission's recommendation causes a distortion to its own detriment only, the provisions of Article 101 shall not apply.*

Title III: Social policy

Chapter 1: SOCIAL PROVISIONS

ART. 117

Member States hereby agree upon the necessity to promote improvement of the living and working conditions of labour so as to permit the equalisation of such conditions in an upward direction.

They consider that such a development will result not only from the functioning of the Common Market which will favour the harmonisation of social systems, but also from the procedures provided for under this Treaty and from the approximation of legislative and administrative provisions.

ART. 118

Without prejudice to the other provisions of this Treaty and in conformity with its general objectives, it shall be the aim of the Commission to promote close collaboration between Member States in the social field, particularly in matters relating to:

- *—employment,*
- *—labour legislation and working conditions,*
- *—occupational and continuation training,*
- *—social security,*
- *—protection against ocupational accidents and diseases,*
- *—industrial hygiene,*
- *—the law as to trade unions and collective bargaining between employers and workers.*

For this purpose, the Commission shall act in close contact with Member States by means of studies, the issuing of opinions, and the organising of consultations both on problems arising at the national level and on those of concern to international organizations.

Before issuing the opinions provided for under this Article, the Commission shall consult the Economic and Social Committee.

ART. 119

Each Member State shall in the course of the first stage ensure and subsequently maintain the application of the principle of equal remuneration for equal work as between men and women workers.

For the purposes of this Article, remuneration shall mean the ordinary basic or minimum wage or salary and any additional emoluments whatsoever payable directly or indirectly, whether in cash or in kind, by the employer to the worker and arising out of the worker's employment.

Equal remuneration without discrimination based on sex means:

(a) that remuneration for the same work at piece-rates shall be

calculated on the basis of the same unit of measurement; and

(b) *that remuneration for work at time-rates shall be the same for the same job.*

ART. 120

Member States shall endeavour to maintain the existing equivalence of paid holiday schemes.

ART. 121

The Council, acting by means of a unanimous vote after consulting the Economic and Social Committee, may assign to the Commission functions relating to the implementation of common measures, particularly in regard to the social security of the migrant workers referred to in Articles 48 to 51 inclusive.

ART. 122

The Commission shall, in its annual report to the Assembly, include a special chapter on the development of the social situation within the Community.

The Assembly may invite the Commission to draw up reports on special problems concerning the social situation.

Chapter 2: THE EUROPEAN SOCIAL FUND

ART. 123

In order to improve opportunities of employment of workers in the Common Market and thus contribute to raising the standard of living, a European Social Fund shall hereby be established in accordance with the provisions set out below; it shall have the task of promoting within the Community employment facilities and the geographical and ocupational mobility of workers.

ART. 124

The administration of the Fund shall be incumbent on the Commission.

The Commission shall be assisted in this task by a Committee presided over by a member of the Commission and composed of representatives of governments, trade unions and employers' associations.

ART. 125

1. *At the request of a Member State, the Fund shall, within the the framework of the rules provided for in Article 127, cover 50 per cent of expenses incurred after the entry into force of this Treaty by that State or by a body under public law for the purpose of:*

(a) *ensuring productive re-employment of workers by means of:*
—occupational retraining,
—resettlement allowances; and

(b) *granting aids for the benefit of workers whose employment is temporarily reduced or wholly or partly suspended as a result of the conversion of their enterprise to other productions, in order that they may maintain the same wage-level pending their full re-employment.*

2. *The assistance granted by the Fund towards the cost of occupational retraining shall be conditional upon the impossibility of employing the unemployed workers otherwise than in a new occupation and upon their having been in productive employment for a period of at least six months in the occupation for which they have been retrained.*

The assistance granted in respect of resettlement allowances shall be conditional upon the unemployed workers having been obliged to change their residence within the Community and upon their having been in productive employment for a period of at least six months in their new place of residence.

The assistance given for the benefit of workers in cases where an enterprise is converted shall be subject to the following conditions:

(a) *that the workers concerned have again been fully employed in that enterprise for a period of at least six months;*

(b) *that the government concerned has previously submitted a plan, drawn up by such enterprise, for its conversion and for the financing thereof; and*

(c) *that the Commission has given its prior approval to such conversion plan.*

ART. 126

At the expiry of the transitional period, the Council, on the basis of an opinion of the Commission and after the Economic and Social Committee and the Assembly have been consulted, may:

(a) *acting by means of a qualified majority vote, rule that all or part of the assistance referred to in Article 125 shall no longer be granted; or*

(b) *acting by means of a unanimous vote, determine the new tasks which may be entrusted to the Fund within the framework of its mandate as defined in Article 123.*

ART. 127

On a proposal of the Commission and after the Economic and Social Committee and the Assembly have been consulted, the Council, acting by means of a qualified majority vote, shall lay down the provisions necessary for the implementation of Articles 124 to 126 inclusive; in particular, it shall fix details concerning the conditions under which the assistance of the Fund shall be granted in accordance with the terms of Article 125 and also concerning the categories of enterprises whose workers shall benefit from the aids provided for in Article 125, paragraph 1(b).

ART. 128

The Council shall, on a proposal of the Commission and after the Economic and Social Committee has been consulted, establish general principles for the implementation of a common policy of occupational training capable of contributing to the harmonious development both of national economics and of the Common Market.

Part V: Institutions of the Community

The Economic and Social Committee

ART. 193

There shall hereby be established an Economic and Social Committee with consultative status.

The Committee shall be composed of representatives of the various categories of economic and social life, in particular, representatives of producers, agriculturists, transport operators, workers, merchants, artisans, the liberal professions and of the general interest.

ART. 194

The number of members of the Committee shall be fixed as follows:

Belgium	12
Germany	24
France	24
Italy	24
Luxembourg	5
Netherlands	12

The members of the Committee shall be appointed for a term of four years by the Council acting by means of a unanimous vote. This term shall be renewable.

The members of the Committee shall be appointed in their personal capacity and shall not be bound by any mandatory instructions.

ART. 195

1. With a view to the appointment of the members of the Committee, each Member State shall send to the Council a list con-

taining twice as many candidates as there are seats allotted to its nationals.

The Committee shall be composed in such a manner as to secure adequate representation of the different categories of economic and social life.

2. *The Council shall consult the Commission. It may obtain the opinion of European organisations representing the various economic and social sectors concerned in the activities of the Community.*

ART. 196

The Committee shall appoint from among its members its chairman and officers for a term of two years.

It shall adopt its rules of procedure and shall submit them for approval to the Council which shall act by means of a unanimous vote.

The Committee shall be convened by its chairman at the request of the Council or of the Commission.

ART. 197

The Committee shall include specialized sections for the main fields covered by this Treaty.

It shall contain, in particular, an agricultural section and a transport section, which are the subject of special provisions included in the Titles relating to agriculture and transport.

These specialized sections shall operate within the framework of the general competence of the Committee. They may not be consulted independently of the Committee.

Sub-committees may also be established within the Committee in order to prepare, in specific matters or fields, draft opinions to be submitted to the Committee for consideration.

The rules of procedure shall determine the particulars of the composition of, and the rules of competence concerning, the specialized sections and sub-committees.

ART. 198

The Committee shall be consulted by the Council or by the Commission in the cases provided for in this Treaty. The Committee may be consulted by these institutions in all cases in which they deem it appropriate.

The Council or the Commission shall, if it considers it necessary, lay down for the submission by the Committee of its opinion a time-limit which may not be less then ten days after the communication has been addressed to the chairman for this purpose. If on the expiry of such time-limit, an opinion has not been submitted, the Council or the Commission may proceed without it.

The opinion of the Committee and that of the specialized section, together with a record of the deliberations, shall be transmitted to the Council and to the Commission.

XIV. THE WAGE PROBLEM IN THE TREATY ON THE COMMON MARKET

1. Wages as international cost factor

The amalgamation of countries which hitherto have been separated from each other by economic frontiers creates the problem of the differences between the structures of costs. If competition between the various countries within the framework of a given industry is intended initially to be based on equal starting conditions all elements of distortion should, of necessity, be avoided. They may appear in the form of differences in taxation, differences in the subsidization of certain industries, between industrial regulations, and so on.

There are in addition differences in costs which are not caused by artificial or governmental intervention but by differences between the degree of scarcity and/or output of the individual factors of production. This applies, amongst others, to the rate of interest on capital and also to wages. A good trade union organization in one particular country might produce a certain increase in the share of the working population in the national income. However, differences between the average wage levels could nevertheless persist, even if trade union organizations were equally strong in the different countries. This is essentially also due to the relationship between supply and demand.

It would be erroneous to believe that the elimination of economic frontiers and/or customs barriers between the various countries in itself would mean that the lower wage level of one country would result in its goods being offered more cheaply and that thereby a certain "wage dumping" would be caused. This conception of "unfair competition" by the countries with a low level of wages and working conditions which is supposed to cause an undercutting of the producers of the remaining countries by some sort of export based on starvation wages and that as a further consequence the market would be controlled by the countries with a low level of wages is wrong, at least in its absolute form.

In the first instance, we always have to consider the output per unit of wages. The costs per working hour are admittedly low in a number of countries, particularly in southern Europe. If, however, the cost of labour is related to the unit of output, i.e., the individual product, a completely different picture will frequently result. It will be found that the cost of labour related to the unit of output is disproportionally high despite the comparatively low hourly wages. This also explains the fact that highly industrialized countries with a high level of wages can compete more than successfully with the domestic manufacturers when offering certain products in which their superior productivity is expressed even within the markets of under-developed countries with a very low wage level.

A comparison of hourly wages, for instance, within Europe shows that the average wage in Scandinavian countries and in Switzerland (calculated in Swiss francs) is more than twice as high as that in the Netherlands and Austria and almost three times as high as Italian wages. Nevertheless the products of many industries of these countries with a high level of wages are cheaper than those of the southern European countries.

These differences in productivity may partly be explained by the differences in the individual efficiency of the workers. The higher level of training and standard of living of the workers of industrialized countries is of considerable importance in this connection. It will as a rule create better conditions for the acquisition of professional qualifications. Nobody will deny in this connection that the results of the efforts of the trade unions help to create the prerequisites which enable the individual worker to increase his output.

The main reason for the difference between the levels of productivity, however, is the fact that the various countries are very differently equipped with capital goods. The Scandinavian and particularly the American workers are so well equipped with machinery and appliances that they have considerably better possibilities to increase their output than their colleagues in less developed countries. Wages are obviously not the only decisive cost factor. The cost of capital, the existence or lack of raw materials, taxes and social charges, finally problems of location of industry, saving in public and private expenditure by means of concentration of industries (Region of the Ruhr), the traffic position, the density of the network of consumption and traffic and other elements are also of far-reaching importance. Comparatively low wages and working conditions may contribute towards a certain equalization of the competitive position in regions where the remaining conditions are unfavourable.

2. The differences between the levels of nominal wages

Any comparison between nominal wages must take into account the differences in the productivity of the various countries. There are, however, several additional factors to be considered.

In the first instance, nominal wages reflect the varying development of the level of prices in the different countries. In countries with an inflationary development the level of prices and wages will be inflated. This applies particularly to those countries where full utilization of labour and capital investments has been the rule for a comparatively long period. The higher level of nominal wages of Sweden and Great Britain as well as of Switzerand and (although to a lesser extent) that of France may be partly explained by these inflationary tendencies.

The wage policy may also exert a considerable influence on nominal wages. Wherever wages may only fluctuate between certain definite limits due to arrangements between governments, employers, and unions and where prices are controlled the result may be completely different from the position which would exist if no controls were imposed.

Generally speaking, wages in Europe tend to decrease the further one goes south. This fact also reflects the degree of economic development of the different countries. Those which are largely agricultural have as a rule a certain "reserve" of unemployed or at least not fully employed people. This is particularly the case in Italy, Spain, and certain Balkan countries. In those countries workers leave agricultural occupations in order to try to find better conditions in industry; on the other hand, the increasing modernization of agriculture causes redundancy. The consequence of this development is a depression of wages and working conditions in industry whereby the already existing differences between the nominal wages in comparison with other countries become even more accentuated.

Countries with a comparatively high level of nominal wages show a tendency to save labour by the use of modern machinery and plant provided a favourable rate of interest enables them to do so. Although this increases the already existing lead of the wealthier countries so far as their output is concerned it may, on the other hand, cause a danger of unemployment due to rationalization and consequently a depression of the level of nominal wages.

According to a number of surveys of more recent date the margin between the costs of labour in various countries appears to have been considerably reduced between 1949 and 1956. There are indications that this development may be ascribed to a number of factors, including the influence of the Treaty on the European Community for Coal and Steel. This applies only partly to France, where a far-reaching adaptation to the more expensive countries has taken place. Whereas France, in accordance with Table I (see below) still occupied the eighth position among the nine enumerated States at the end of 1949, she had already advanced to third place at the end of 1956.

TABLE I.

INDEX FIGURES FOR THE COST OF ONE WORKING HOUR IN THE YEARS 1949, 1955 AND 1956

FRANCE = 100

December 1949		October 1955		October 1956	
U.S.A. ...	490	U.S.A. ...	344	U.S.A. ...	343
Switzerland ...	170	Switzerland ...	132	Sweden ...	133
Sweden ...	150	Sweden ...	101	France ...	100
Belgium ...	129	France ...	100	Switzerland ...	98
Great Britain ...	124	Belgium ...	94	Belgium ...	95
Federal Germany	110	Great Britain ...	93	Great Britain ...	93
Italy ...	106	Federal Germany	88	Federal Germany	89
France ...	100	Italy ...	74	Italy ...	75
Netherlands ...	84	Netherlands ...	58	Netherlands ...	58

Source: Figures of the French Institut National de la statistique et des études économiques, published in the journal "Etudes et conjoncture", August 1957, page 882.

Whenever nominal wages are considered due note should be taken of the fact that their conversion on the basis of the rates of exchange only insufficiently expresses the actual differences between the standard of living because these rates of exchange frequently tend to be on the high side. So far as industrial enterprises are concerned however, the factors which matter to them in competition are not the actual differences between the real income but the costs of wages because they influence the price of their products.

3. Differences between social duties and benefits

The social policy of the various countries comprises two different types of benefits.

There are on the one hand grants, mainly in countries where the level of wages is, generally speaking, particularly low. The insufficient development of the economy coupled with frequently divided and therefore less effective trade union movements caused various States to grant special subsidies through the channel of social welfare in order to complement the inadequate wages and thereby counteract extremist political tendencies. These benefits have become known under various headings, such as family allowances, marriage grants, children's and birth allowances, reductions of the price of bread, cheaper housing, reduced contributions to insurances, removal grants, and so on.

On the other hand there are contributions in the nature of assurance premiums which are provided by the State and industry and which correspond to benefits granted to the part of the population covered by these contributions. Certain changes have taken place in this field in recent years inasmuch as social provisions applicable to the entire population have been introduced to cover the consequences of illness, old age, invalidity, and other contingencies of life.

The influence of social duties and benefits on the nominal income varies very considerably in the individual European countries. This may be illustrated by Table II which is based on a survey of the I.L.O. In accordance with that table the average real wage in France in 1954, for instance, only amounted to approximately 60 per cent of the average Swedish wage and 75 per cent of the average Swiss wage. The average in Germany was even slightly lower than in France and Italian hourly wages amounted to no more than 37 per cent of those paid in Sweden. A completely different picture however, results when social contributions and benefits are also taken into account. It will then be seen that France only lags behind by approximately 8 per cent and that there is only a difference of approximately 25 per cent between the average nominal hourly wage in Germany and Italy and that in Switzerland. Arrangements for the payment of wages during holidays and vacations have also been included in these calculations.

TABLE II.

AVERAGE HOURLY INCOME, SOCIAL EXPENDITURE AND COST OF VACATION IN EUROPE

Country	Average hourly wage 1954 In Sw. frs.	Index Switzerland = 100	Compulsory social insurance in per cent of wages as on 1 Jan. 1956	Cost of days of vacation and holidays in per cent of wages 1952 and 1953 as charge on employer	Wages plus compulsory social benefits plus cost of vacations and holidays (Index Switzerland = 100)
Sweden	3.33	130	2.2	6.0	128
Finland	2.81	110	6.0	4.5	111
Denmark ...	2.67	104	7.5	6.5	108
Norway ...	2.58	101	3.7	6.0	101
Switzerland ...	2.56	100	3.9	6.0	100
Great Britain ...	2.26	88	2.7	6.0	88
Belgium ...	1.91	75	17.7	11.6	88
France	1.88	73	29.8	7.2	92
Federal Germany	1.74	68	11.7	9.8	75
Ireland	1.53	60	1.7	4.6	58
Netherlands ...	1.38	54	19.0	7.2	62
Austria	1.30	51	19.0	10.1	60
Italy	1.22	48	53.5	14.2	73

From: Social Aspects of European Economic Cooperation, Report by a Group of Experts (Ohlin-Report), Geneva, International Labour Office, 1956, page 33.

Later surveys of the French Institute of Statistics and Economic Studies, however, seem to indicate that the effects of social policy have become less expressed in recent years. This appears to be due to the fact that in a number of countries the social expenditure of which previously amounted to a smaller percentage of the cost of labour has considerably increased since 1948/1949.

In view of the fact that the basis of calculation used in Table III differs from that used in Table II a comparison of the figures will meet with certain difficulties. Such comparison is furthermore less important than the fact that the figures shown against each year in Table III demonstrates a tendency towards a harmonization. It is also worth noting that social costs are of reduced importance for the enterprise, the products and wages wherever old age, social and health insurance are largely covered from public funds; this applies particularly to Great Britain.

TABLE III.

SOCIAL EXPENDITURE

	Germany	Belgium	France	Italy	Netherlands	Great Britain	Sweden
In per cent of the gross wages bill							
1948 ...	33.2	24.6	28.5	56.2	23.1	—	—
1949 ...	34.0	24.0	—	58.7	23.2	—	—
1950 ...	—	24.1	29.4	57.3	23.4	—	—
1951 ...	37.0	25.4	—	—	24.9	—	—
1952 ...	—	27.8	31.7	65.2	25.3	—	—
1953 ...	—	28.2	—	69.3	26.0	—	—
1954 ...	—	28.8	33.3	70.3	—	—	—
1955 ...	41.0	30.0	—	70.8	—	—	—
1956 ...	—	30.5	—	71.2	—	—	—
In per cent of the real wages							
1955 ...	41.0	29.0	42.7	63.5	28.6	8.6	15.3
1956 ...	41.0	29.4	41.9	63.5	28.6	8.4	15.3

According to : "Etudes et conjoncture", published by the French Institut National de la statistique et des études économiques, August 1957, pp. 868-883.

4. Differences in the level of real wages

The differences in the cost of living are not only influenced by the level of nominal wages calculated in accordance with the official rate of exchange. In many instances the differences between the nominal wages are reduced by virtue of the already mentioned fact that in countries with a high level of nominal wages the level of prices is frequently also relatively high.

Any comparison between the levels of real wages in the different countries meets with a number of difficulties. The differences in the general way of living are so great that it is hardly possible to establish a general scheme of consumption, even if only for western European countries. The Italians, for instance, consume more fruit and wine, the northern Europeans, however, more meat and fats. This results in rather curious consequences. It has, for instance, been calculated that an Italian would have to pay considerably more if he wanted to buy the same goods in Britain which in his home country are everyday consumer goods. Conversely, an Englishman would have to pay more in Italy than at home if he wanted to buy precisely the same things which he likes to see on his table in London. Consequently, an Italian who lives in England would, despite a relatively higher wage, be worse off in certain circumstances or at any rate no better off than if he had remained in Italy where he earned a lower wage. It is due to these differences in the structure of consumption that there is very little statistical material available concerning comparisons between real incomes. Only in recent years was an attempt made within the framework of the European Community for Coal and Steel to establish a uniform scheme of consumption based on consumer habits in the different countries of the Community which have in each instance been expressed in the form of a "shopping basket" of consumer goods. These shopping baskets were evaluated separately for the different industries of the Community for Coal and Steel, namely for coal mining, iron-ore mining, as well as the iron and steel industry. In view of the fact that the survey concerning coal mining is based on consumer habits of more than one million workers its figures may be considered as fairly representative.

In the following Table IV an attempt is made to arrive at an international comparison of real wages based on "parities of consumers' expenditure" and the hourly wages in the different countries. The "parities of consumers' expenditure" are intended to show how much one has to spend, for instance in D.Ms. or French francs in order to purchase the same quantity of goods which may be obtained for 100 Belgian francs in Belgium.

TABLE IV.

COMPARISON BETWEEN REAL INCOMES OF WORKERS

		Average hourly incomes in manufacturing industries[1] (1956)	Parity of consumers' expenditure[2] in bituminous coal-mining (1954)		Real wage[3] in B.frs. (1956)	Index of real wages[4] (Belgium = 100) (1956)
Germany ...	D.M.	1.76	8.43	=	20.90	81
Belgium ...	B.frs.	25.80	100	=	25.80	100
France ...	F.frs.	177	767	=	23.10	90
Italy	Lire	201	1,306	=	15.30	59
Netherlands ...	Gulden	1.32	6.40	=	20.60	80

[1] According to Etudes et conjoncture, No. 8, August 1957, page 879.

[2] According to E.C.S.C. Comparisions between real incomes of workers of the industries of the Community, Luxembourg 1956.

[3] Calculated by dividing [1] by [2]. The insignificant changes in the parities of consumers' expenditure between 1954-56 have been neglected due to lack of exact statistics.

[4] Index calculated according to [3].

It is rather interesting to note in which way conditions, as illustrated above, have changed in the course of the past thirty years. A useful basis for this purpose is a survey of the I.L.O. for the year 1927. A comparison (Table V) shows that the position has relatively deteriorated since then in Germany and the Netherlands. This may be due to various reasons.

In the 'twenties a strong trade union movement existed in Germany. The dictatorship which followed, including the wage freeze which it imposed, stopped any development for more than a decade. It has apparently not been possible to recover from this setback entirely during the post-war years.

In the Netherlands, an extensive wage and price control was imposed mainly for the benefit of the export trade. The industrial peace which was thereby obtained has influenced real wages to a certain extent.

In France the workers were able—precisely during the years of the setback in Germany—to obtain a considerable increase in their real wages and of their share in the national income (Léon Blum period). Although the effectiveness of trade union action has decreased during the post-war period, a system of social benefits granted by the State nevertheless maintains real wages in France at a relatively high level. If, however, only the actual wages—without social benefits—are considered, their purchasing power appears to have lagged behind considerably, at least until 1953.

We reiterate that the competitive position is not reflected in the level of real wages. How much each worker is able to buy with his wage in his country also depends, and by no means in the last resort, on import policy and on the attitude of the government towards agriculture. Where the domestic market is deprived of cheap products of the world market by protectionist measures, an improvement of real wages will, from a long-term point of view only be possible within certain limits, even if the development of wages in general were favourable.

TABLE V.

INDEX FIGURES OF REAL WAGES 1954/56

COMPARED WITH 1928

FRANCE = 100

	Index of real wages[1] 1956	Coal-miner underground (married with two children)[2] 1954	Coal-miner underground (married without children)[2] 1954	Steel worker (married with two children)[2] 1954	Steel worker (married without children) 1954	Index of real wages according to I.L.O. statistics for capital cities[3] 1928
	According to figures of the E.C.S.C.[2] *"Shopping basket"*					
Federal Germany	90	76	90	84	103	114
France	100	100	100	100	100	100
Italy	65	67	74	79	90	78
Netherlands ...	89	99	114	83	96	149
Belgium ...	111	103	113	108	124	91

[1] In accordance with Table IV.

[2] From "Die Arbeitereinkommen" E.C.S.C. (see Table IV) pp. 136—138.

[3] From International Labour Review. Oct./Nov. 1928, pp. 658—9. The comparison only applies to capital cities and merely takes into account the cost of food; the differences become slightly reduced if rents are also taken into consideration.

5. Limits of harmonization

The opinion has frequently been expressed that the common market could only function if there were a certain harmonization of the financial burdens resulting from the costs of labour and the social policy. This contention is basically erroneous. As already explained, differences between the nominal wages of the various countries reflect differences in the level of productivity which in turn result from diffences in the availability of capital goods, raw material, manpower, and so on. It is not possible as yet to determine to what extent these differences may be eliminated in the course of time. The natural lead which certain countries enjoy in regard to the productivity of certain industries, however, cannot and should not be abolished artificially. One of the main prerequisites of the intensification of the exchange of goods and services which is one of the objectives of the Common Market will be the particularly favourable position of certain industries of individual countries which will consequently be in a position to exert a stronger influence within the framework of an efficient division of functions. This will, however, only be possible if the more advantageous cost positions which justify such division of functions are allowed to fully come into play.

It may, on the other hand, be maintained that a certain harmonization will always become necessary whenever additional costs arise due to artificial interventions, if a distortion of the conditions of competition is to be avoided. This applies particularly in the field of taxation and social charges which are imposed on the enterprises by the authorities, in other words, more or less from outside. A common denominator could be found for these artificial cost factors, provided they do not result from natural market conditions.

It is frequently claimed that from a long-term point of view taxes and social charges ought to be harmonized, in principle, so far as their repercussions on prices are concerned. If this meant that the proportion of social charges expressed as a percentage of the average hourly wage as, for instance, obtaining in Italy would now have to be aimed at in other countries as well, it would be a rather debatable contention. The social expenditure of the Italian enterprise compares with social benefits of the State which the latter has to grant precisely because the general level of wages is inadequate. In those countries where a wage based on output is paid which corresponds to the requirements of the worker there is no reason for artificially extending the share which the social wage represents in relation to the total wage. This is all the more relevant in view of the fact that the "social wage" paid by the State is largely outside the control of the trade unions. Neither does it appear to be possible to find, comparatively quickly, a common denominator for the working conditions negotiated by means of collective agreements. The harmonization of these cost factors does consequently in our opinion not appear to be feasible, at least for the time being. The British solution of a general social and health service which divorces the problem of social expenditure from the sphere of the individual industrial undertaking is, we believe, more useful because welfare thereby becomes an obligation on the community.

Other solutions must consequently be attempted. The standard of living of the working population must be improved by reinforcing the action of trade unions in countries where it has not reached its full efficiency due to organizational weakness or bad economic conditions. This procedure would also level out to a certain extent the differences which exist in the cost of labour.

The "White Book of Messina", in which these problems were

considered early in 1956, suggests that a number of differences in costs and in the rate of taxation could be eliminated to a certain extent by the adaptation of rates of exchange to the parity of purchasing power. There is no doubt much to be said for this idea. The White Book, however, also pointed out that special measures of harmonization would become necessary wherever individual industries which are subject to heavy burdens are particularly adversely affected. This may be correct where social measures are financed by contributions of industry which are related to the amount of wages. In such cases industries with production costs with a high labour content are more affected than those where the capital content is higher. Conversely, if social security measures are predominantly financed from the budget and consequently represent a charge on the entire economy, enterprises the production costs of which contain a larger labour content will be in a more favourable position. The question will therefore arise whether a certain equalization of the systems of social contributions ought not to be attempted. Furthermore, an adaptation of the rates of exchange would have to be aimed at because a reduction of the value of a currency of a country with a particularly high rate of taxation tends to balance conditions in international competition.

6. Possible repercussions of the Common Market on the wage level of the different countries

Among the more important repercussions of an economic integration on the level of wages are movements of migration. For a number of reasons, however, such movements on the part of the workers are not likely to be too extensive. If one disregards the refugee problem, only approximately one million workers have migrated to other countries in Europe since the end of the war, despite many facilities which have been granted in this connection. The majority of them were Italians. In France the immigration of African labour has additionally to be considered. Skilled labour shows a rather reluctant attitude towards migration. According to the experience of the European Community for Coal and Steel the possibilities of initiating large-scale migrations are limited, even if considerable assistance towards adaptation were made available.

A certain harmonization of the level of nominal wages would more likely result from competition between the enterprises, in view of the fact that an adaptation of the prices of goods and services will in the long run also assist an adaptation of the prices of the factors of production. It will therefore be seen that the question of productivity becomes one of major importance.

The differences in nominal and real wages in Western Europe are to a considerable extent due to the differences in productivity. A certain harmonization could be obtained in this connection if major movements of capital were brought about. A better equipment of southern European industries could lead to a considerable increase in wages and a raising of the standard of living.

It is, however, quite possible that not even capital would migrate to the extent which would be required in order to bring about an equalization of the differences in productivity and consequently in real wages. An industrialist will tend to invest capital wherever he may expect large profits. This is mainly possible in regions which have already reached a high standard of economic development. The danger will consequently arise of precisely the intensively industrialized regions attracting new investments and capital. Regions which have hitherto always suffered from a scarcity of capital consequently might suffer from an even more acute scarcity of capital after the creation of the Common Market and therefore the differences in productivity would become rather more accentuated than mitigated. The differences in the development of Southern and Northern Italy—after the unification during the last century—seem to indicate that such danger may be very real indeed. On the other hand, the reverse may also be the case. Industrialists could be tempted to invest capital wherever production could be carried out at relatively low cost due to a low level of nominal wages. This has been the case in the United States where light industries have for many years shown a distinct tendency to migrate from the north towards the south.

Which of these opposed tendencies will finally prevail will not only depend on the differences between nominal wages and the influence of social benefits. Factors which have a bearing on the location of industries, such as traffic connections, prices of real estate, possibilities of recruiting skilled labour, raw materials, and so on, are equally important. A direction of capital might have to be resorted to in order to arrive at an intensification of investments in regions with a scarcity of capital. The improvement of power supplies in under-developed areas by means of a policy of controlled investment may also acquire a certain importance.

XV. THE WAGE PROBLEM IN THE TRANSPORT INDUSTRY

As we have explained in the present study the level of nominal wages is influenced to a major extent by the level of productivity, apart from a number of additional factors. The level of productivity may in turn become a cost factor of major importance in international competition. Whether and to what extent special problems arise for the transport industry in view of those facts is by no means definite.

In certain circles the opinion is held that differences in costs affect services in the same way as they affect the production of goods. In both cases the structure of costs of one country was in immediate contact with that of another country. It was therefore in connection with the employment of comparatively cheap labour of no importance whether such cheap labour was employed in the production of goods for export or in international transport. It was always a case of selling comparatively cheap labour abroad. Low costs of production, however, were in similar instances equivalent to offers at a low price which would lead to a more intensified international division of labour and consequently to an overall improvement of the standard of living.

The other opinion is based on the assumption that the differences in productivity are not as clearly expressed in connection with services, and particularly in the transport industry, as they are in the remaining branches of the economy. If one assumed, however, that the efficiency of vehicles of equal standard was everywhere the same and that the drivers met identical requirements, the nominal wage and the costs of the working conditions would acquire an increased importance. A transport operation covering the same distance with an identical lorry would cost less in Sicily than in Belgium. That difference would not disappear even if other cost factors such as the prices of fuel, taxes and so on were considerably higher in Sicily than in Belgium.

If transport based on low costs of labour now passed a frontier and entered a country with comparatively high costs of labour, the structure of costs of the "cheap labour" was carried into that of the "expensive" labour. That process was not mitigated to the same extent as in the case of export goods where the distance in itself, due to the costs of transport, already fulfilled the function of a protective duty. In contrast with the export of goods it was in this instance not a product but, as it were, the factory itself which was exported whereby one element of its structure of costs, namely the wage, did not originate in the internal conditions of the country of destination but was a part of the structure of costs of the foreign country. According to the same opinion the actual differences would become even more accentuated if the vehicle operated by "cheap" labour did not return empty to the frontier. In contrast with, for instance, a machine used in industrial production which, having reached its destination, would begin to operate within the framework of the structure of costs of the country of destination, further transport operations would still be carried out at the lower costs of the country of origin. This applied in the case of return freight as well as, and perhaps primarily, in the case of cabotage within a common transport market. Consequently special problems of wage policy would arise within the transport industry.

This way of reasoning is not accepted by those representing the other opinion.

The differences in the structures of costs are reduced to a certain extent by the payment of expenses to the transport worker as a compensation for the higher expenses of the country to which his journey takes him. An additional factor to be considered in this connection is the migration within the transport industry which will presumably exceed the general average and will consequently contribute to a gradual harmonization. It will furthermore be necessary to aim without further delay at a harmonization of the wage and cost conditions within the transport industry in order to secure an undisturbed realization of the common transport market at the earliest possible date.

For an evaluation of these problems three factors are of major importance: the labour costs in relation to the total cost; then the differences between the levels of nominal wages in the various countries; and finally the level of productivity. It is unfortunately very difficult to make available or to calculate comparative figures related to the transport industry. Our following tables are consequently by no means perfect; they may, however, nevertheless serve a fairly useful purpose.

Table VI illustrates the share of the costs of labour in the total cost. These data have only been available for a few countries. So far as road transport is concerned, the comparison becomes difficult for the further reason that the figures are not related to exactly the same size of vehicle. Furthermore the full coverage of track costs has also frequently not been taken into account in this connection. If it had been, the figures, for instance, relating to Italy would be lower. So far as the railways are concerned, it is certain that the proportion of the costs of labour is considerably higher than in the case of road transport. Again, that proportion is higher for short distance goods transport than for long distance goods transport.

Consequently general wage increases imply a higher charge on railways than on long-distance road haulage. An equalization could only be achieved by a correspondingly more rapid development of productivity of the railways which may hardly be expected despite the congested roads which impede the development of road transport to an ever-increasing extent.

TABLE VI.

SHARE OF LABOUR COSTS IN TOTAL COSTS

Country	Total average road transport	Railways' share of wages in total operating costs	Road short-haul goods traffic (and/or below 5 ton)	Road long-distance goods traffic (and/or over 5 ton)
Italy 1954[1]	34	—	32(44)[4]	21
Denmark 1953[1] ...	36	—	—	—
Sweden 1956[1]	—	—	35	23
Sweden 1950[2]	—	67[3]	—	—
West Germany 1950[1] ...	—	—	26	25
Great Britain 1950[1] ...	32	—	—	—
Belgium 1950[2]	—	70[3]	—	—
France 1950[2]	—	60[3]	—	—
U.S.A. 1950[2]	48	61[3]	—	—
Canada 1950[2]	37	60[3]	—	—
Europe 1954[1] ... ... (total average)	—	54	—	—

[1] According to Economic Commission for Europe: costs and efficiency of road and rail transport. Economic Survey of Europe, Chapter VI, 1956.

[2] According to International Labour Organization, Coordination of Transport: Labour Problems, Geneva 1951, page 148.

[3] These figures appear rather high. They represent the share of wages in operating costs, i.e., the total costs minus interest on operating capital and all costs of operations not connected with goods or passenger traffic. The actual proportion will very likely correspond to each of the figures minus 10 or more points.

[4] Below 2 tons.

In the first and second columns of Table VII an attempt has been made to produce a summary of the different nominal wages in road transport and on the railways of the various countries. So far as the figures based on the calculations of Colin Clark relating to road transport are concerned, it should be noted that they represent general averages of short- and long-distance traffic. The two columns may not be compared with each other. Each of them separately shows that there are very great differences between nominal wage levels in all enumerated countries. If one neglects the data concerning the U.S.A. on the one hand and Spain, Portugal, and Yugoslavia on the other, that spread is reduced in the case of the railways to 55 points, whereas it still amounts to 155 points in road transport. This difference is last but by no means least a consequence of the existence of generally strong trade union organizations of railwaymen in the free countries whose successses also enabled the weaker countries to make up some leeway. The result of this comparison is also likely to be influenced by the greater similarity of the organizational structure and the differences in the methods of survey. A comparison between these data and those contained in Table II on page 117 shows that the differences between the nominal wages of the various countries are greater in road transport than in relation to the general average.

TABLE VII.

WAGES AND LABOUR COSTS IN EUROPEAN TRANSPORT INDUSTRY 1953-54

Country (Belgium = 100)	Wages Index per number of workers		Cost of labour related to traffic units 1 passenger/km. = 1 ton/km.
	Road haulage 1953	Railways 1954	Railways 1954
U.S.A.	560	320	80
Sweden	240	130	110
Switzerland	190	—	—
France	—	125	130
Denmark	135	110	140
Norway	—	110	200
Belgium	100	100	100
Great Britain	105	—	—
Germany	100	90	110
Netherlands	85	80	50
Italy	—	80	70
Austria	70	75	80
Greece	—	75	130
Spain	30	45	80
Portugal	55	40	100
Yugoslavia	—	30	30

Col. 1. Calculated according to: Colin Clark, The Conditions of Economic Progress, 3rd Edition, London, 1957, pp. 526—531.

Cols. 2 and 3. According to E.C.E. Economic Survey of Europe, Chapter V, Geneva 1956.

So far as productivity is concerned the relevant figures have only been calculated for the railways. Our figures in column 3 of Table VII are round figures, in view of the fact that they can only represent approximate values. In connection with the definition of the unit of traffic one passenger kilometre has been assumed to equal one ton-kilometre. Although certain doubts may arise as to the effectiveness of this method of calculation it has nevertheless become a general rule in connection with a number of statistics of this type.

Any comparison should take into account the fact that in thinly-populated, mountainous countries like Norway and Greece the costs per traffic unit could not easily be reduced even if there were a high level of productivity because the aforementioned difficulties stand in the way of the effectiveness of rationalization measures.

XVI. BASIC PROBLEMS OF TRANSPORT POLICY WITHIN THE COMMON MARKET

1. Differences between the transport policies of the various countries and their repercussions on competition

The elements which distort competition exist in the different countries to a varying extent. Attempts should therefore be made within the framework of integration to harmonize such distortions so far as possible.

Efforts in this respect have been impeded by the considerable differences between the structures of transport policy of the various countries.

There are, for instance, States where one branch of the transport industry may undercut another without any difficulty, whereas in other countries attempts have been made to restrict such developments by tariff obligations applicable to all branches of the transport industry and/or a quota system applicable to competition-intensive branches of the transport industry. So long as such differences exist, however, an abolition of frontiers will very likely mean that relatively unfettered branches of the transport industry of one country will exert a disturbing influence on the balanced structure of transport of another country and consequently also influence its structure of wages which has in the meantime begun to function well. A harmonization of measures in the field of transport policy within the entire territory of the Common Market appears to be the most likely prerequisite for the avoidance of such disturbances of competition. The following brief survey of the structures of transport policy of the different countries will serve as an explanation of the main differences which exist:

(*a*) Germany

In the German Federal Republic the Minister of Transport determines the basic principles of a common policy of all branches of the transport industry and is also responsible for the determination of fixed rates. In this connection the interests of the community play an important role and are taken into consideration by means of special exceptional rates by which, among others, the areas affected by the Iron Curtain, economically undeveloped parts of Western Germany and certain industries which are in need of support are being assisted. Other exceptional rates are intended as a counter-measure against competition by inland navigation.

Federal Railways and long-distance road transport are linked together in the field of rates and fares. The long-distance scale favours transport over longer distances to a certain extent. So far as wagon-loads are concerned there is an *ad valorem* scale consisting of seven classes in which connection the Federal Railways advocate a reduction to four (in actual practice, three).* The differential between the classes

* Five ad-valorem classes for wagon-load traffic were introduced on 1st February 1958.

A to G over a distance of 200 kilometres has up to now been approximately 100 to 46. The volume of single consignments is taken into consideration in the secondary classes. Their rates, however (as short distance rates in general) do not even remotely approach the cost differentials. Up to the present time there have only been in existence rates for five, ten, and fifteen tons.†

So far as long distance road haulage is concerned there is apart from the obligation to apply certain rates also a strict limitation of the number of vehicles.

In short distance road haulage the application of maximum rates is compulsory. There is a general obligation to publish rates.

Fixed rates are also applicable to inland navigation and they are determined by the freight commissions for the various river and canal areas. East of Hamm there are also restrictions on foreign traffic.

Coastal navigation is only open to German ships.

(*b*) **Netherlands**

Conversely, the transport industry of the Netherlands is mainly governed by the principles of private enterprise. The rates structure is largely adapted to average costs and market conditions and social considerations do not play any important role in this connection. Rates and fares of the railways are in actual practice fixed by the railways themselves. The relevant powers of the authorities are limited to the approval of rates. So far as wagon-loads are concerned an *ad valorem* scale consisting of four classes is applicable; the differential amounts to approximately 100 to 66. Rates take into consideration the volume of individual consignments whereby a certain protection against goods which take up an unduly large space in relation to their weight is afforded.

A large volume of the traffic is carried at special rates. They are in many instances based on agreements with individual traders and need not be published by the railways. The Netherlands Railways have been fully compensated by the State for damage caused as a consequence of war and occupation, which has probably very largely contributed to the fact that the railways of the Netherlands are the only principal railways within the Economic Community which have been continuously showing a profit.

Tramp traffic in inland navigation is subject to maximum and minimum rates which are fixed by the competent authority after consultation with private industry. Scheduled traffic is governed by maximum rates.

So far as road transport is concerned minimum and maximum rates are applicable, particularly in international road transport, which are fixed by the authorities after consultation with the private carriers.

† A main class for 20 tons was also introduced on the railways on 1st February 1958.

A system of concessions and permits is applicable in inland navigation and road transport but it does not correspond to a proper application of quotas.

Coastal navigation is not subject to any restrictions.

(*c*) **Belgium**

Rate policy in Belgium is largely based on average costs. The railway company fixes its own rates and the government is merely entitled to raise objections. There are eight *ad valorem* classes for wagon-loads. In the case of small parcels there is a sort of door-to-door rates structure. These goods are, however, carried at express rates. There is one main class for twenty tons. In foreign traffic reductions are granted for groups of wagons.

No supervision of rates exists in Belgian road transport, neither at national nor at international level. There is merely a system of concessions and permits which involves the application of a certain quota.

Freight rates in inland navigation are also completely free of restrictions in traffic across frontiers (with the exception of an agreement on quantities of sand and gravel for the Netherlands) whereas at national level the application of rates which are based on average costs is compulsory. Consignments are allocated by freight agencies.

Coastal navigation is a preserve of the national flag.

(*d*) **Luxembourg**

The rates structure of the Luxembourg railways differs from that of the Belgian railways in that it only consists of six *ad valorem* classes. Road transport is not subject to any compulsory application of rates. Special agreements are not permissible.

(*e*) **France**

The rate system of France partly takes into consideration certain social elements, e.g., by means of reduced fares in commuters' traffic. So far as goods transport is concerned certain elements of rates between traffic centres (distributive rates) have been applied whereby the better cost position of railways on main lines comes into play. A number of secondary lines have been closed, particularly for passenger traffic.

Railway rates are subject to the obligation to be published. Rates over short distances up to 100 kilometres are exceptionally high.

The system of rates is based on *ad valorem* scales and the principle of optimization of capacity of the wagons and comprises four series (classes). The differential between the highest class for wagon-loads and coal is 100 : 42. So far as normal rates are concerned there is a minimum and maximum rate in each class with a differential of 2.5 per cent. The total differential between the highest maximum rate and the lowest minimum rate is up to 30 per cent.

The railways thus apply maximum and minimum rates, which

take into account average costs and are fixed within the framework of an index calculation of the individual stations (coefficient of frequency of a line). There are also quantity rebates for complete trains or groups of wagons and a sort of goodwill-rebate for faithful customers, as well as contract rates for maintaining traffic which are not pubished.

Scheduled road passenger traffic is subject to approved rates. In goods traffic provisions have been made for the introduction of maximum and minimum rates containing certain elements of the *ad-valorem* scale; the actual rate is then to be fixed by agreement between customer and carrier within the limits determined by these maximum and minimum rates. It is felt, however, that it would be rather difficult to supervise the application of such rates.

So far as road transport is concerned there is furthermore a special system involving the issue of tickets which, as it were, enable an allocation of capacity to be carried out.

Inland navigation is subject to a system of compulsory rates and compulsory allocation of freight by means of special freight agencies as well as an allocation of ships.

Coastal navigation is a preserve of the national flag.

(*f*) **Italy**

The railways have to apply fixed rates. The principle of geographical equality of rates is being maintained. The level of rates is deliberately kept low by the government (ratio between rates index and cost-of-living index 50 : 100) in order to take into account the geographical position (long distances from north to south). More extensive increases in rates only took place in 1957, mainly for products of the coal and steel industries. The deficit of the railways is covered by the State. This is in accordance with the Italian conception intended to balance the costs of tracks. The distance scale tapers comparatively steeply. Short distance rates up to 100 kilometres are high. The secondary classes are 20 to 32 per cent above the rates for 15 tons. Within the *ad valorem* scale the differential between the classes 41 to 85 is 100 : 30. Special rates for 15 tons are only applied from class 66 onwards. The differential between the classes 66 to 85 over distances of 200 kilometres corresponds to 100: 62.5. The obligation to carry wagon-loads and also small parcels is only applicable to a distance of at least 30 kilometres. Special rates agreed with certain traders are not published

There is a system of exceptional rates similar to that applicable in Germany which, however, is not a consequence of competition by inland navigation (which, apart from Northern Italy, is practically non-existent) but rather by coastal navigation. Foodstuffs are carried at rates which are considerably below cost level.

A forwarding agency owned by the railways is largely responsible for the transport of small parcels and collective consignments as well as collection and delivery.

There is no compulsory application of rates in road transport. Relations between rail and road transport are not subject to any regulating measures.

Coastal navigation is largely reserved to the national flag. Special agreements have been concluded with different countries.

———o———

In addition to the above-mentioned differences there are also very considerable differences in taxation of road transport and charges levied on inland navigation in the various countries. In accordance with statistics of the I.R.U. the average tax imposed on a 32-ton lorry-trailer combination in Western Germany, France, and Italy based on a yearly performance of 60,000 kms. in 1956 amounted to approximately DM. 15,000 whereas carriers in the Benelux countries operating under identical conditions were only subject to taxation to the extent of approximately DM. 6,000.

Although similar differences in taxation may, so far as trade in general is concerned, be partly counter-balanced via the rates of exchange this is not possible in the transport industry if the transport service is rendered abroad and the invoice made out on foreign currency.

It would therefore appear to be indispensable to adopt certain harmonization measures in order to prevent the development of differences in remunerativeness and consequently between wage levels in the transport industry, if a common transport market is to become a reality.

2. Possibilities of changes in location of industries and their repercussions on transport policy

The abolition of the existing elements of a number of rate systems which take into account the interests of the community (i.e. geographical equalization of rates, particularly in Germany and Italy and certain special rates in a number of countries) may entail a number of changes in the location of industries. This would particularly be the case if such elements were abolished in an inharmonious way and also in the case of areas situated at some distance from the centres of economic activity of the various countries. As a consequence the frequency of traffic in remoter areas might be even further reduced which would cause additional costs particularly so far as branches of the transport industry are concerned which are subject to the obligation to carry and to maintain operation, i.e. mainly the railways.

The only possibility of a certain harmonization of these developments could consist in a number of areas situated near the inland frontiers of the Common Market (for instance, Eifel and Ardennes) now becoming more closely connected with the centres of economic activity of other countries if and when economic frontiers are abolished. The latter areas would consequently be less affected by the above mentioned changes. Similarly increased transit traffic in the direction of marginal areas may be conducive to a certain general improvement of the position.

There is on the other hand also a possibility of certain changes in the location of industries occurring within the entire territory of the European Economic Community in the course of development of the Common Market. This could, for instance, apply to re-locations of the centres of the coal and steel industries in the direction of the ore deposits in Lorraine or even Central France. There could also be tendencies to locate heavy industries in the sea ports, for instance, at the Rhine Estuary. Whereas the first-named possibility would counteract trends towards an excessive concentration and might in certain circumstances be conducive to a more equally distributed volume of traffic for all branches of the transport industry as a whole, the latter possibility could lead to a very noticeable reduction of the volume of important inland traffic in bulk commodities.

The repercussions of such changes in location on the branches of the transport industry will mainly have to be tested according to the extent to which any losses can be compensated by a general stepping-up of the exchange of commodities and possibly by a general economic boom within the Common Market.

3. Mobility of labour and the equalizing function of the Social Fund

We have already pointed out that only a limited increase in the willingness of labour to migrate may be expected as a consequence of the creation of the Common Market. Tourist traffic too is not likely to increase to any noticeable extent. On the other hand the greater facility in crossing frontiers might be conducive to an increase in the volume of commuters' traffic between the various countries. Such traffic, insofar as it is carried at social rates may lead to additional elements of disparity arising particularly in railway passenger transport.

Migrations of labour are, however, particularly important for the transport industry if foreign transport labour is employed within the territory of other states on behalf of foreign transport enterprises. These problems which have already been discussed earlier on are likely to become very urgent indeed. They give rise to a number of questions of harmonization of wage and working conditions as well as of social conditions.

If the developments to which we have also referred in earlier chapters were indeed to take place and lead rather to an increased concentration of areas with a surplus of capital than to a harmonization through migration of capital, certain pockets of unemployment might result, particularly in the case of a general economic recession which would also affect transport workers.

The Social Fund provided for in the Treaty on the European Economic Community only offers limited possibilities to counteract similar developments. In view of the fact that rehabilitation and re-settlement grants are envisaged, particularly in connection with the repercussions of the creation of the Common Market, the danger of discrimination between two types of unemployed may arise in case of an economic recession (those who have become unemployed through the Common Market and receive grants and others who may have lost their employment because of the general recession and receive no rehabilitation or re-settlement grants). This problem may acquire topical importance within the transport sector in view of the fact that the transport industry is immediately and to a particularly noticeable extent affected by adverse economic developments. The only efficient means to avoid such pockets of unemployment appear to be structurally planned regulating measures in the field of investment. It will above all be necessary to adopt measures in order to prevent certain countries trying to mitigate the repercussion of possible economic crises by simply deporting to their home countries foreign workers who have become unemployed. There can be no question of any policy of exporting unemployed labour within the framework of the community.

4. Integration, coordination and common transport policy

We should like to stress the fact that the creation of a common transport market is an indispensable prerequisite for an exchange of commodities. Consequently attention should be paid in the very first instance to measures which may be conducive to a coordination of the different systems of coordination in the various countries. The measures envisaged in Part One of this study would therefore have to be unified within the framework of the European Economic Community. As a sequel to such a "super-coordination", a common transport policy would then have to be determined. This would be extremely difficult if the different countries were to maintain their full sovereignty. In that respect the creation of a joint transport commission for Europe which the trade unions too had advocated in the past could have been fairly helpful. There can at any rate be no doubt whatever that any harmonization will meet considerable difficulties if the different measures adopted in the field of national transport policy were maintained and divergent tendencies at national level were to be further encouraged.

Within the framework of this problem of a general coordination of measures in the field of transport policy numerous problems arise in connection with the various sectors and branches of the transport industry which are discussed in greater detail in the following chapters.

XVII. THE QUESTION OF THE INCLUSION OF A FREE TRADE AREA

The question which will have to be answered whenever the repercussions of the free trade area within the transport sector are to be considered is: will it be possible without any particular difficulty to "liberalize" the supply of services within the territory of other states as easily as the exchange of goods?

In view of the fact that this question involves coordinating measures in the various countries, an immediate reply in the affirmative will hardly be possible. If it is, however, to be answered in the affirmative, a mutual adaptation of coordinating measures would become necessary for the inland transport of the countries concerned at international level as well as within the actual sector of integration. As a consequence the difficulties in connection with such adaptation would be largely the same as in the case of the countries of the European Economic Community.

Differences may, for instance, arise between the German Federal Republic and Austria in view of the fact that tendencies to reduce the *ad valorem* scale have lately become apparent in Germany whereas Austria appears to favour the maintenance and extension (1952) of the *ad valorem* scale.

Other countries which may be involved in inland traffic with the territory of the Treaty in view of their geographical position as, for instance, Denmark and Switzerland do not impose any far-reaching obligations on road transport as applied in Germany and envisaged in France.

The large majority of the countries which are eligible for a free trade area (Great Britain, Sweden, Norway, Portugal, and Greece) cannot however, for geographical reasons be of any immediate importance so far as the problems of the railways, road transport, and inland navigation are concerned. In these instances, the free exchange of services is primarily restricted to shipping and civil aviation, i.e., to branches of the transport industry which are excluded from the field of competence of the Treaty on the Common Market.

If they were to be included, problems would arise primarily in connection with coordination and/or the elimination of national priorities within the field of coastal navigation and civil aviation. So far as navigation on the high seas is concerned the question of joint protective measures against unfair competition by countries which do not belong to the free trade area could become topical, particularly with regard to the Panlibhonco ships. In connection with the problems of primary and food industries certain regulating measures within the fishing industry (subsidized prices, fishing schemes, etc.) might also have to be discussed with a view to their coordination; these discussions would, however, only indirectly affect the transport industry.

XVIII. THE RAILWAYS WITHIN THE COMMON MARKET

1. Joint efforts

The monopoly of the railways and the similarity of gauge forced the European railways quite early in their history to agree on a close cooperation in the technical field. In order to facilitate international railway traffic across frontiers the administrations simply could not avoid arriving at agreements concerning the gauge, loading and unloading installations, as well as the dimensions of the rolling stock used in traffic across the frontiers. Due to the pressure of competition by road transport and inland navigation, this cooperation has subsequently been extended beyond the initial objective of only unifying technical data and it now also covers matters of general importance for the transport industry. In that respect, we are admittedly still in the initial stages of a lengthy process. It has, for instance, taken centuries to arrive at a uniform loading gauge for the rolling stock used in international rail transport. The harmonization of the tariff structures will very likely meet even greater difficulties. The natural differences between the opposing interests are in this connection of an even more immediate impact than in the purely technical field, due to the varying competitive position as well as the differences between the structures of the economies. It will only be possible to gradually reduce these differences in the course of integration of the Western European economic area.

The unification and improvement of the conditions of the construction and operation of railways, so far as the international traffic is concerned, is particularly due to the efforts of the U.I.C. (International Railway Union). That organization, which was founded in 1922 represents today a network of over 310,000 kms. (approximately 193,000 miles) with a personnel of 4,000,000 and a yearly operational performance of 240 milliards of passenger/kms. and 270 milliards of ton/kms.

The executive bodies of the U.I.C. are the Central Clearing House at Brussels (B.C.C.), the Office for Research and Experiments (O.R.E.), at Utrecht, the Information Centre of the European Railways (C.I.C.E.) at Rome, and the International Railway Documentation Bureau (B.D.C.) at Paris. In order to avoid duplication of work and differences of opinion with other international railway organizations an agreement— the "Agreement dated 1st January 1951" — was concluded in 1951 between the U.I.C. and the "International Committee for Railway Transport" (I.T.K.), International Carriage and Van Union (R.I.C.), the International Wagon Union (R.I.V.), as well as the European Goods Timetable Conference.

In accordance with this agreement, each organization maintains its complete autonomy, but the U.I.C. is responsible for the cooperation between these organizations. Since the conclusion of that agreement the U.I.C. has been representing all international railway unions at the different international organizations.

In the year 1951, an agreement was concluded between France and Western Germany concerning the common use of goods wagons in traffic between these countries (Europ Agreement). By virtue of this arrangement the joint unrestricted use of goods wagons by different railway administrations has for the first time become a reality without changing anything so far as the ownership rights are concerned. The agreement between Germany and France was extended in March 1953 to eight further Western European railways and confirmed in its revised form by all ten administrations (DB, EdS, OeBB, SNCB, DSB, SNCF, FS, CFL, NS, SBB/CFF) on 1st October 1955.

Up to the end of 1956, the European rolling stock consisted of 167,000 goods wagons. With effect from 1958, goods wagons intended to be included in European rolling stock must fulfil certain uniform conditions regarding loading capacity, wheel base, and service weight, which have been fixed by the U.I.C.

The unification of the design of goods wagons is furthermore being encouraged by the agreement for the financing of the purchase of rolling stock (Eurofima) which was signed on the occasion of the Conference of European Ministers of Transport (C.E.M.T.) by fourteen member states at Berne 1955; in this instance, too, however, efforts to arrive at a unification are subject to certain limitations. The reasons for the differences in the development of European rolling stock construction are the varying climatic conditions as well as the incidence of important goods to be carried and furthermore special customs and requests of the trading community.

The initial stage of unification in the field of railway transport which has been reached will have to be further developed within the framework of the European Economic Community. The most important objective of a common European rail transport policy, however, appears to be the harmonization of railway tariffs and the introduction of direct international rates in traffic within the European Economic Community. The attainment of these objectives is admittedly a difficult problem, in view of the diverging tariff policies of the railways caused by the differing economic structures of the various countries. These, in turn, depend on the geo-political position (central or marginal position), on the geographical structure (mountains, plains, coasts), on the differences between the wage, financial and monetary policies of the various states as well as on the varying competitive position of the railways.

A unification of tariff systems may consequently only be carried out within the framework of a common economic and transport policy and the concomitant joint coordinating measures.

2. The question of unification of the ad valorem scale

The elaboration of a uniform goods list for the railways of the six countries which are partners to the Treaty appears to be comparatively simple. It becomes, however, more complicated if that list, apart from its inherent purpose, also is intended to stipulate in which *ad valorem* classes the railway administrations have to group the different types of merchandise. So long as differences exist between the economic structures and the competitive position of the various countries, it will not be possible to deprive railway administrations of their freedom to decide on the classification of goods at their discretion without creating major conflicts of interests. A railway which suffers from considerable competition by road haulage so far as the traffic in certain highly-priced goods is concerned will be forced to group these goods for competitive reasons in a comparatively lower *ad valorem* class than another which is not exposed to such competition. Similar conditions apply to goods of lower value in competition with inland navigation.

It is true that it would be easy to arrive at an agreement on the number of the *ad valorem* classes without thereby putting the railways to any noticeable economic disadvantage. Their competitive interest and their special tasks within the framework of economic policy will be safeguarded as long as they are still authorized to determine at their own discretion the differentials between the different classes as well as the overall extent of the entire *ad valorem* scale. Such a sweeping simplification of the *ad valorem* scale would admittedly not be of much use because it could not prevent considerable discrepancies between the rates charged on certain goods even if all railways grouped them within the same *ad valorem* class. It is quite feasible that, for instance, a merchandise of the country A is grouped in Class 3 of the *ad valorem* scale, which, however, so far as the actual rate is concerned, could correspond to Class 2 of country B or Class 4 of country C. A positive unification based on the prerequisite of a uniform grouping of the goods in certain *ad valorem* classes could only be obtained if the scale of rates and the differentials between the different groups of rates were fixed uniformly for all railways. This would, however, be bound to meet with difficulties because of the differences between the competitive position of the various railways and the differences between their economic obligations and their tasks within the field of social policy.

A unification of the *ad valorem* scale is furthermore complicated by the fact that there are railways where the pure *ad valorem* principle is only applied in conjunction with the principle of weight/space ratio, i.e., the grouping of goods in accordance with the space which their weight occupies in a normal goods wagon. This problem could, however, be solved by refraining from applying either the principle of the weight/space ratio or the pure *ad valorem* principle.

3. The question of unification of graduation of rates according to the distance

Generally speaking, freight rates of all European railways consist of two elements: terminal service charges and the actual rail charge.

The *terminal charge* is independent of the distance over which the merchandise is carried and represents a compensation for the transport costs arising at the place of despatch and destination. These so-called fixed costs can obviously differ in accordance with the varying structural basis of the railways. Any unification of terminal charges could therefore, if it is intended to take into account the costing of the railways, only relate to the special services which may be included under terminal services.

So far as the actual *rail charge* is concerned, the different railway administrations apply various methods of calculation of the rate of graduation. The majority of administrations apply tariffs which are reduced in accordance with the distance, whereas others have a system of proportional freight rates based on a structure which is particularly adapted to local requirements. In addition to this tapering-off of the actual rates, the majority of administrations which apply this system also vary the rate of graduation according to the *ad valorem* classes.

So far as the extent of the graduation is concerned, prime costs are no longer of any importance for the majority of administrations, because it is thought that the costs of carrying one ton of merchandise do not always show a proportional decrease with the increasing distance. It could also be possible that the costs increase and/or remain constant, for instance, due to increased shunting operations. The coefficient of graduation has very frequently been governed by considerations of competition with other branches of the transport industry as well as by overall considerations of the economic policy. This may be taken as an indication that the tapering-off of rates according to the distance largely takes into account special commercial interests of the railways which could no longer be safeguarded if any rigid system of unification were applied.

4. The question of unification of the coefficient of surcharges in the secondary classes

The majority of railway administrations of the states which are partners to the treaty apply a system of uniform weight classes of five, ten, and fifteen tons. There are, however, other administrations which have introduced weight classes which vary according to the type of merchandise and also a greater number of weight classes. An agreement concerning the number of weight classes would very likely not be too difficult to obtain.

A unification of the coefficients of surcharges, as for instance applied by the European Community for Coal and Steel, appears, however, more difficult. In this connection, two main aspects have to be considered, namely, prime costs and the competitive position. A unification of the coefficients of surcharges in accordance with prime costs will immediately meet considerable difficulties in view of the fact that there are very noticeable differences between these prime costs due to the varying geographical and economic structure within the European

Economic Community. It should also not be overlooked that the railways, when calculating their coefficients of surcharges, for competitive reasons nowadays attach greater importance to the actual market conditions than to their prime costs. In most European countries the coefficients of surcharges are almost without exception lower than the full prime costs of the secondary classes. Admittedly, in this instance, too, the position varies from one country to the other. Furthermore, the problem of return freights acquires particular importance in the case of radial traffic which is a rather prominent feature of certain railways and implies the necessity of an elastic assessment of the coefficients of surcharges in the secondary classes.

5. The question of unification of the speedy-conveyance scale

At the present time, three types of freight services are uniformly provided by the railways of nearly every country: ordinary freight, fast freight, and express freight. A number of railways furthermore use the so-called accelerated fast goods trains. There is, however, one noticeable difference, inasmuch as in the majority of countries the consignor is entitled to a free choice between fast and ordinary freight for all types of merchandise. In a number of countries he can, on the other hand, not avail himself of this choice. He may only apply for despatch by fast goods train if this type of conveyance is not normally provided for a given merchandise. In order to avoid an alternative solution of the question of a uniform system, the consignor could in principle be allowed free choice between both types of despatch, whereby, however, an additional charge could be levied for all traffic for which no despatch by fast freight is provided. It will be more difficult to arrive at a unification of the differentials between these types of conveyance. Once again, the differences between the prime costs and competitive interests of the railways seem to deprive attempts at unification of any real purpose.

Calculations of delivery times differentiate between the time required for the handling of goods at the terminal and time of conveyance. Certain railways fix a uniform delivery time for all distances. Although it might be possible to obtain a unification of the regulations governing terminal handling times for all railways, this would not be possible in the case of the times of conveyance which may show very considerable differences in accordance with the density of traffic on various lines. If any unifications were attempted in this field, it would have to be based on the possibilities of conveyance along the least frequented route, a procedure which would cause negative repercussions on the competitive position of the railways compared with other branches of the transport industry.

The above considerations show that the success of any efforts aiming at unification of European railway goods tariffs always tends to be jeopardized if such efforts entail economic disadvantages for certain railways by not sufficiently taking into consideration the actual competitive position and the economic structure within the field of activity of these railways. The above-mentioned tariff elements are the main instruments of competition at the disposal of the railways in their struggle against the competition by road haulage and inland navigation. These elements would lose their inherent purpose if a rigid unification deprived the railway administrations of their freedom of action. In the case of a coordination of the entire Western European inland transport, these aspects would only lose some of their importance if the coordination were carried out purely on the basis of a planned economy with complete exclusion of the play of supply and demand. In view of the conceptions of economic policy underlying the treaty, this assumption does, however, hardly appear to be justified. Finally, the fact that a unification of the tariff systems presupposes a unification of the legal status of the railways concerned should also not be overlooked. All this proves that the harmonization of railway tariffs may only be considered as a long-term aim. In this connection the advantages which the European Economic Community would derive from such unification should be carefully weighed against the disadvantages which it could entail for certain railways.

6. The question of elimination of broken freight rates at the frontiers (through rates)

On looking back, it may be stated that the introduction of direct rates in accordance with the method developed by the High Authority of the European Community for Coal and Steel was not a happy solution. The determination of a uniform coefficient of graduation in traffic across frontiers causes considerable work and restricts the freedom of the railways to determine their tariff policy at their discretion. The arguments against this procedure are not only justified by the prime costs of the railways and the losses of revenue which are presumably caused by the application of such through rates (as borne out by the example of the European Community of Coal and Steel) but are equally justified by the competition between the railways. All these objections have no doubt contributed to the fact that the Treaty on the Common Market, in contrast to that on the European Community for Coal and Steel, does not contain any provision for through rates in traffic across frontiers.

It has been thought that the additional costs arising from frontier formalities would justify a freight rate broken at the frontier. In our opinion, however, this argument is not valid. Apart from the fact that the work of the "L.I.M. Conference" already aims at a considerable simplification and acceleration of frontier formalities by means of joint railway stations and completion of customs formalities by inland custom offices before despatch or delivery of the goods across the frontier, the actual problem of the direct through rates would still remain unsolved, even if all existing frontier formalities were completely abolished. Neither do the costs arising from shunting a wagon when transferring from one railway administration to another justify a breakage in freight rates at the frontier. These shunting operations are in no way different from the making up of trains in national traffic which does not stand in the way of the application of through rates. The argument that the functional relation between the tapering-off of costs and the distance of conveyance is open to doubt and that the actual transport charges in reality hardly show any tendency to taper off does also not militate in principle against an abolition of the freight rates broken at the frontier. If prime costs were taken into account to a larger extent when calculating the rate of graduation of freight rates the only consequence would be more reduced rate of tapering-off than is nowadays generally applied.

The main difficulties which stand in the way of through rates stem from the losses of revenue which they would involve for the railways in traffic across frontiers. Graduated through rates would make transport in international goods traffic cheaper. This would admittedly stimulate trade within Europe but at the same time cause a deterioration of the financial position of the railways. Whether these losses could be compensated by a general increase of railway tariffs appears rather debatable in view of the competition between the railways and other branches of the transport industry. The railways themselves have for obvious reasons always viewed any raising of the level of their tariffs with a certain scepticism, because it is precisely the cheapness of conveyance which acts in favour of the railways in competition with road haulage which frequently is technically superior. From a long-term point of view a certain compensation of the losses of revenue could indeed be effected by the stimulation of trade subsequent to the introduction of direct tariffs. Assuming that an increase of the general willingness to have goods conveyed is possible, it must equally be presumed that the general economic trend will continue in an upwards direction. The European Community for Coal and Steel came into existence under the auspices of an economic boom. Whether this could be said in respect of the Common Market is rather doubtful.

It is, however, not only the through rates in traffic across frontiers which may entail losses in receipts of the railways. Any reduction of the cost of transport—as in this instance—must of necessity entail a diversion of the main international and national streams of rail traffic, which in turn would cause a distribution of railway revenue differing from that prevailing today. This problem will admittedly lose some of its importance with the increasing organizational fusion of the railways of the countries of the European Economic Community. It will, however, remain as long as this process of fusion has not been initiated.

We believe that from a long-term point of view direct rates in rail traffic of the European Economic Community will prove just as unavoidable as they were unavoidable in national rail traffic. We are, however, of the opinion that great prudence ought to be applied when introducing them. Direct rates should initially be applied wherever international flows of traffic already exist along the so-called main routes of international rail traffic. This might offer an opportunity for the introduction of European tariffs between principal centres which might cause a return to the railways of part of the traffic which they had lost in international traffic to road and inland navigation.

7. Survey of rail traffic within the European Economic Community

The rail network of the countries of the European Economic Community extends over more than 110,000 kms. This corresponds to approximately one-third of the entire European rail network excluding the Soviet Union. The lowest density of the network within the Economic Community can be found in Italy, with 6.3 kms. per 100 sq. kms. The highest rail density arises in Belgium with 24 kms. per 100 sq. kms. In comparison the average density of the entire European rail network (excluding the Soviet Union) is 6.1 kms. per 100 sq. kms. Within the European Economic Community, the railways carry annually approximately three thousand million passengers and 650,000,000 tons of goods which corresponds to a yearly operational performance of more than a hundred thousand million passenger/kms. and over 130,000,000 ton/kms.

RAILWAYS WITHIN THE EUROPEAN ECONOMIC COMMUNITY

Country	*Total rail length Km.*	*Kms. per 100 sq. kms.*	*Electrified kms.*
Belgium	7,280	23.8	1,995
German Federal Republic	37,050	15.0	2,134
France	43,270	7.9	5,515
Italy	18,970	6.3	7,475
Luxembourg	465	19.8	70
Netherlands	3,178	9.3	1,400

From: Jahrbuch des Eisenbahnwesens, Folge 8. Carl Röhrig Verlag, Darmstadt 1957.

XIX. ROAD TRANSPORT WITHIN THE COMMON MARKET

1. International and inter-governmental regulations

Whereas the railways are already subject to national and international regulations, thanks to efforts extending over decades, various attempts are still being made in road transport to arrive at a mutual adaptation and unification of regulations. An important contributory factor to this state of affairs is not only the multiplicity of rival interests along the common road but also the multiplicity of vehicles and undertakings.

There is hardly a better way to describe the wide field covered by arrangements made since the second world war or to be made than a brief enumeration of the efforts within the framework of the Economic Commission for Europe (E.C.E.), the competent regional organization of the United Nations. During the ten years of its existence the Inland Transport Committee of the E.C.E. has comprehensively dealt with all aspects of road transport. Thus main roads and their characteristics have been defined and a code of signalization, traffic regulations and equipment of vehicles has been elaborated. Agreements have been reached or at least attempted on road signals and markings, level crossings, weights and measures, questions of insurance and customs and, last but by no means least, for a variety of aspects of tourist traffic on the road. The questions of temporary entry of commercial vehicles and, as a corollary, the problems of dual taxation have also been settled by agreements.

The present study is mainly concerned with the efforts which have been undertaken with a view to regulating road transport including commercial transport and transport on own account. The E.C.E. has devoted a considerable part of its work to this field. The legal foundations of an international contract for carriage of goods on the road were created by an agreement of the year 1956, which may be considered as counterpart of the C.I.M. agreement of the railways. It is intended to create a similar basis for passenger and removal transport.

So far as traffic across frontiers is concerned, a provisional agreement was concluded in 1949 which simplifies customs formalities for vehicles and goods and which has become known by the letters T.I.R. (Transports internationaux routiers). This agreement is intended to be finalized in 1958.

The problems of taxation have been the subject of two agreements concluded in 1956. They make provision for exemption from transport tax for coaches and lorries in transit. The same committee has considered the discriminations which stem from transport taxes.

Whereas road transport at national level is in the main regulated by a system of licences, permits and quotas an analogous solution at international level still remains to be found. It is true that the agreements concerning the freedom of the road concluded at Geneva in 1947 provided for the following freedoms: for international circular coach tours; for journeys beginning at an air or sea port and terminating there; and finally, for the transit of goods. No agreement has, however, been reached on the type and the granting of permits and licences. This question is closely connected with the problem of coordination and it is therefore obvious that governments are afraid of committing themselves too far by an international agreement of this type. It should, at any rate, be noted that national tourist traffic in all countries is reserved to the national enterprises.

As a consequence of the ineffectiveness of the agreement on the freedom of the road and in view of the need to regulate to a certain extent the continuously increasing international road traffic, agreement was reached within the E.C.E. in 1954 after protracted negotiations on what is without any doubt the most important convention between the E.C.E. States, namely the *"General Agreement on Economic Regulations for International Road Transport"*. It consists of a Set of Rules which determines the obligations to which carriers and vehicles in international traffic are subject, and of the General Agreement itself, which introduces this annex. The Agreement regulates the entire international goods and passenger traffic including under the first heading transport on own account. The contracting States may by virtue of the Agreement only ask carriers entering their territory to abide by international agreements. Any existing national regulations which might be more severe may be imposed on these carriers. The Agreement is of considerable importance for the workers, because their social conditions are regulated in a special Annex A. This part was prepared in cooperation with the International Labour Organization. It stipulates, *inter alia*, that no more than nine hours may be spent at the wheel within a 24-hour period and that the driver must be granted a regular rest period of ten hours (exceptionally eight hours) outside the vehicle.

For the time being, this very comprehensive agreement has two main disadvantages: although it has been signed by eleven states, amongst them five states of the European Economic Community, it has still not been ratified. At least five ratifications are, however, required if the agreement is to become valid at all. Apart from Greece and Norway, only France and Italy had ratified the agreement up to the end of the year 1957. The main reasons for this reticence appear to be the distinct differences in the matter of rates policy which also has to be covered by the agreement. In Germany fixed rates are applied after a certain distance. France may introduce a system of minimum and maximum rates which have to be fixed or approved by the authorities. A similar system but with certain modifications has been applied in the Netherlands in international transport for some considerable time. In other countries there is complete freedom in the matter of tariff policy with the occasional exception of maximum limits which are fixed by the price control authority. Sometimes this freedom is dependent on the obligation to publicize rates or at least to notify the competent authorities. The differences of opinion appear to be caused by the extent to which these existing regulations ought to be considered within the framework of the agreement. We are of the opinion that the absolute freedom to determine tariffs belongs to the past and that the obligation to publicize rates is, apart from any special agreement, the minimum on which any international regulations could be based.

The second disadvantage consists in the fact that no agreement has so far been reached concerning the procedure governing permits or licences, i.e., the admission to international traffic. It is intended to solve this problem by a special agreement. Until then the individual contracting States will endeavour to settle details by means of bilateral or multilateral agreements.

The European Community for Coal and Steel initially did not attach any undue importance to the question of road haulage. In view of the fact, however, that the traffic of products of the Community on the road has increased, solutions are being sought which take into account the provisions of Article 60 (Remuneration) and 70 (Conditions of transport) of the Treaty on the European Community for Coal and Steel. It is thus intended to create regulations governing the determination of maximum and minimum rates and the supervision of their application.

In view of the fact that the Geneva agreements are in many instances still ineffectual and that the efforts of the European Community for Coal and Steel have not yet been finalized, bilateral agreements in international road transport between individual states are for the time being still of the greatest importance. Although there are differences in the basic principles underlying these agreements it is nevertheless true that generally speaking traffic across frontiers is dependent on a system of permits and licences which, as a rule, are granted within a certain annual quota. If one examines the road haulage between the member-states of the European Economic Community, it will be found that traffic between Germany and the Netherlands is by far the most important.

The mutually agreed quota of vehicles which is laid down in a provisional agreement amounts to 1,050 lorries which carried more than two and a half million tons of goods in the year 1956. In comparison traffic between Denmark and Germany, which is also considered one of the most intensive traffics within Europe, only reached a volume of half a million tons, carried by 280 vehicles.

The volume of goods carried by road between the Netherlands and Germany is, however, not the only reason for the importance of that traffic. It merits particular attention in this connection because the transport policies of these two countries are based on fundamentally different principles. If, therefore, a comprehensive agreement should finally be reached as a result of the negotiations which have been going on for years, it may be assumed that the basic principles underlying such an agreement might be applied to a certain extent also by other countries and perhaps also within the European Economic Community.

2. The question of the unification of regulations governing permits and licences

In Part One, Article 3, of the Treaty on the Common Market the elimination of the obstacles which stand in the way of a free exchange of persons, services and capital between member-states is mentioned and the introduction of a common policy in the field of transport envisaged.

The system of permits and licences which exists in the various states of the European Economic Community no doubt represents an obstacle to the free exchange of services in road transport. It is consequently safe to assume that the regulations to be adopted must aim at an elimination of these obstacles. One factor which must, however, not be overlooked in this connection is that the systems of permits and licences have been introduced in the very interest of road transport and with a view to its coordination with other branches of the transport industry. Similar efforts within the European Economic Community may no doubt be expected.

We feel that it would be feasible to find a solution in this case by the application of the regulations on the elimination of quantitative restrictions in goods traffic to the restrictions in road transport in accordance with their inherent meaning. This would mean that each Member-State would convert the bilateral quota into a bulk quota. At a later stage a gradual extension of these allocations could be effected as long as this would appear desirable within the framework of a common transport policy and consequently also within the framework of co-ordination with other branches of the transport industry.

So far as the internal traffic of the various states is concerned, it may be assumed that generally speaking a numerical restriction of commercial vehicles in road traffic will be maintained. Unlimited access of foreign carriers to the inland traffic of the various states may consequently not be expected. A complete opening up of the home markets would jeopardize any success of coordinating measures and would lead to ruinous competition in wide fields of the Treaty. A gradual opening up of the home market could in our opinion only be effected by allotting a certain percentage of the national quota of vehicles in inland transport to foreign vehicles. That percentage should be capable of revision in order to allow for an adaptation to the market conditions prevailing at any given time. It might be feasible to issue some sort of international licence to foreign carriers who are admitted.

As a prerequisite of the introduction of such a licence the conditions governing their issue in the various states would in the first instance have to be unified. In the further stages on the way towards a free exchange of services the introduction of an E.E.C. licence may become important inasmuch as it would in actual fact initiate a bulk allocation of vehicles at the level of the Common Market.

The solution of the problem of permits for scheduled transport of goods and passengers becomes more difficult in view of the fact that it involves preferential treatment for the owner of such permit. This obviously proves that a suitable basis could only be created by means of a common transport policy and, as a corollary, a common policy of coordination.

Administrative measures which require unification are, however, not the only obstacle which stands in the way of a free exchange of services. For the time being there are also obstacles of a technical nature. According to an international convention, 1949, each State is at liberty to determine maximum weights and measurements of vehicles, but the minimum weights and measures which must be applied on certain main roads are determined by that convention. The German Federal Republic, however, does not adhere to these standards and is also not prepared to join any regional agreement to which the Benelux countries and France already adhere within the framework of the Economic Community. It is obviously believed within the Federal Republic that these standards are no longer adequate in view of the density of traffic and no longer correspond to the requirements of safety and an unhindered flow of traffic and that an attempt ought to be made to reduce the costs of construction and maintenance of the roads also by the expedient of reducing measures, weights, and payloads. In our opinion these differences could only be eliminated by neutral experts who would examine conditions in view of the present-day situation in road transport, assisted by representatives of all interested parties, and whose recommendations would be supported by comprehensive arguments.

3. The question of the right of establishment

In accordance with Articles 52 to 58, the text of which we have reproduced earlier on, it is intended to gradually eliminate existing restrictions of the right of establishment within the Economic Community. The inherent mobility of road transport and its special structure may justify the assumption that the right of establishment may assume particular importance in this connection.

The main obstacles which stand in the way of freedom of establishment are in the first instance the national quotas and the differences between the technical requirements. On the other hand the efforts aiming at an extension of this right are supported by the fact that road transport undertakings in one part of the territory of the Treaty are offered better competitive positions than in the remaining part. This applies particularly to fiscal policy, but also to wages and social conditions.

It is obvious that each road transport undertaking will very carefully weigh the advantages and disadvantages of a transfer to the territory of another E.E.C. country in order to finally settle wherever the best possibilities for the future development of business are expected to exist. It is on the other hand equally obvious that there can be no question of complete freedom of choice of establishment as long as there is any danger of coordinating measures already adopted in the field of transport being jeopardized. A necessary prerequisite of any positive development is consequently in this instance, too, a common transport policy of the contracting States.

4. Problems of tariff structure

We have already pointed out in the first part of this chapter the fundamental differences between the tariff policies of the contracting States. There are some with more or less far-reaching tariff obligations and others where there is complete freedom in the field of rates policy. These divergencies create special problems also for road transport.

What should be the total charge for a transport operation which covers the territory of a country with obligatory rates as well as that of another country where they are non-existent? Should there be complete freedom to base charges on the total distance? In that case a longer transport operation might in certain circumstances be cheaper than a shorter one covering the sector within the country with obligatory rates. This could, however, result in distortions of the conditions of competition.

If on the other hand the total charge were based on the controlled rate for one part of the distance and complemented for the remaining part by a freely agreed freight rate, an element of insecurity with regard to transport charges would be introduced in traffic across frontiers, which would be detrimental to road transport as a whole.

These problems, too, can only be solved within the framework of a common transport policy. Its prerequisites are again contained in Article 75 of the Treaty.

In accordance with the stipulations of that article it is the duty of the Council of Ministers of the Community to establish common regulations for international traffic. Such regulations would have to take into account the requirements of coordination of the branches of the transport industry as well as the importance which is attached to transport as a public service.

There can be, in our opinion, no question of trying to find a solution on the basis of rate freedom in road transport. The inevitable consequence of such a policy would be that railways and inland navigation of the contracting States would be granted the same freedom. The consequences of such a development can easily be imagined.

Conversely, the introduction of fixed rates would no doubt meet with the determined opposition of those countries which hitherto have applied a liberal tariff policy. The objections of those countries that fixed rates largely eliminate competition and could not reflect the actual market development cannot be entirely dismissed.

The measures which would have to be adopted are consequently likely to be based on the determination of maximum and minimum rates. An arrangement of this type would be more elastic and allow for a price structure which would be adapted to competition as well as to the conditions prevailing at any given time.

A system of determination of tariffs of this type would nevertheless meet with considerable difficulties so far as its application is concerned. The question consequently arises whether it would be useful to introduce an *ad valorem* scale in road transport as has been the rule in rail traffic. If so, should maximum and minimum rates be applicable to each class and who should determine them? Should the rates take into account the overall economic interests of the community or should it be possible to determine them freely within the fixed limits?

We need not go into further details of this question. It has been thoroughly examined in the first part of our report in connection with the problems of coordination. We believe at any rate that a common policy in this question too can only be arrived at if at the same time the problems of coordination in the entire field covered by the Treaty are treated as priority.

5. Development of road transport since 1930

The importance of the problems with which the authorities of the Economic Community are faced in the field of road transport alone may already be judged by the quantitative importance of road traffic. We feel that the time is not very far away when the economic problems of road transport will no longer in the first instance originate from competition but to a far greater extent from its increase, particularly with regard to individual means of transport and consequently the congestion on the existing roads.

The following table was published in the "Economic Survey of Europe", United Nations, Geneva, 1957:

Increase in the number of motor vehicles in Europe since 1930

Country [a]	Number of motor vehicles [b] in 1954			Passenger cars				Lorries			
	Total	Per 1,000 inhabitants		1930	1938	1954	1956[c]	1930	1938	1954	1956[c]
	Thousands	Passenger cars	Lorries	*Index 1950 = 100*							
Sweden	652	74	14	41	62	211	292	54	67	127	132
France [d]	3,802	62	25	62	113	149	191	49	52	133	152
Great Britain ..	4,372	63	22	47	85	135	169	38	54	118	131
Luxembourg .	25	60	19	63	81	195	—	65	98	135	—
Belgium	570	45	19	36	56	146	183	40	55	171	130
Denmark ...	289	44	21	67	92	164	205	53	69	159	175
Switzerland ..	286	50	8	41	51	162	210	39	54	127	154
Norway	188	32	25	42	95	166	208	38	69	159	179
Ireland	162	42	13	41	58	134	148	36	43	150	173
West Germany	2,158	30	13	54	138	270	394	27	60	164	164
Netherlands ..	345	22	10	49	72	158	228	58	63	123	152
Finland	121	17	11	91	—	264	388	35	—	147	199
Italy	1,091	16	7	60	96	222	318	34	37	148	172
Austria	152	13	8	34	55	179	370	32	31	123	147
Portugal	131	10	4	40	—	143	—	19	—	140	—
Spain	218	4	3	160	—	131	—	58	—	122	—
Turkey	66	1	1	22	34	213	244[e]	21	25	197	223[e]
Yugoslavia ..	34	1	1	142	212	177	219	18	27	126	137

Sources: National statistics.

[a] Based on the number of vehicles in relation to the number of inhabitants.

[b] Excluding motor-cycles.

[c] Estimates.

[d] Unofficial estimates.

[e] 1955.

XX. INLAND NAVIGATION WITHIN THE COMMON MARKET

1. Legal background and possibilities of development

Any examination of the problems of inland navigation within the European Economic Community must be preceded by a consideration of the existing international agreements and treaties as well as the relevant provisions of the Treaty concerning the European Economic Community.

In this connection, the three conventions of Barcelona, 1921, are of fundamental importance. The "Convention and Statute on the regime of navigable waterways of international concern" divides the internationalized rivers into three categories. These categories are governed by the composition of the supervising authority responsible for a given waterway. The Rhine, for instance, is considered an internationalized river of the second degree because its Commission consists of representatives of riparian and non-riparian states.

All parties to the agreement enjoy freedom of navigation on the waterways of all three categories. Nationals of the participating states, as well as goods and flags are treated on a basis of equality. Charges may only be levied as a compensation of actual expenditure incurred in connection with traffic facilities. A provision of particular importance is that concerning cabotage, which stipulates that each riparian state may reserve the right of cabotage.

In a protocol to the agreement, the signatory States pledge themselves to equal treatment of the flags of all States which are partners to the convention on all territorial and non-internationalized waterways so far as import and export traffic is concerned, always, however, reserving their territorial rights and on the condition of reciprocal treatment.

Similarly, transit traffic is treated in a further agreement. All these conventions were ratified by twenty and more states. Even where they have not been ratified, however, they have always been considered as an important pillar of international legislation in the field of inland navigation.

So far as regulations for the Rhine are concerned which is of particular importance within the European Economic Community, the Mannheim Act of Navigation of 17 October 1866 is the instrument of fundamental importance. It has been revised or supplemented several times and thereby adapted to developments. Provided its provisions and those concerning the maintenance of safety are applied, shipping from Basle as far as the open sea is permitted to the vessels of all nations. No charges may be levied on the Rhine and its tributaries in so far as the latter are on the territory of the participating States. Ships trading on the Rhine and their cargoes are treated by the contracting States in all respects on the same basis as their own ships and cargoes.

The decisions of the Central Rhine Commission are reached by an absolute majority. They only become binding for the riparian states, however, if their governments have approved these decisions. This provision is of importance inasmuch as Switzerland is also a riparian state but not part of the European Economic Community.

These traditional provisions are reinforced and extended by the treaty concerning the Economic Community. In accordance with Article 61, the free exchange of transport services, is governed by the provisions concerning transport which we have reproduced earlier on and commented upon. In this connection, particular importance must be attached to Article 75 which for the time being is interpreted in different ways.

2. The present position

Generally speaking, it may be said that the European undertakings which engage in international inland navigation are not subject to any state-imposed restrictions concerning their participation and charging. In actual practice, however, there are certain limitations and restrictions which, as is widely believed, are incompatible with the conceptions underlying a common transport market.

In the traffic with France on waterways other than the Rhine, for instance, there exists a compulsory system of "turns." In accordance with this system all ships engaged in tramp traffic must register for cargoes at a freight exchange. The conclusion of short-term transport contracts with the trading public is governed by the sequence of registration at the freight exchange. Long-term contracts may, however, be concluded without regard to these turns.

The fifty-fifty clause applicable in Belgium to the transport of sand and gravel is another type of restriction.

Although these restrictions imposed by the authorities do not in themselves exert an influence on the freight structure and do not imply any discrimination either, they are nevertheless generally considered as an interference in the free participation in navigation.

The restrictions imposed on free participation on the basis of private contracts, in the form of reserved traffic or based on cartel agreements, are, however, far more drastic.

The transport of coal for Swiss gas works and of coke from the nationalized Dutch mines to Basle for instance is exclusively in the hands of a limited number of shipping companies. The French Federation of Shipping Companies, "Communauté de Navigation Française Rhénane" (C.N.F.R.), is also in many respects protected from foreign competition by similar agreements. Consequently, any international competition, even between undertakings of one and the same nationalities, is frequently eliminated. In actual fact the reserved traffics are nothing but a camouflaged type of false transport on own account because the shipping companies concerned do not restrict themselves to reserved traffic but also participate in the competition of the purely commercial goods transport for the account of third parties.

The influence of cartel agreements and pools is discussed in a special chapter.

If the prerequisites for a free exchange of transport services are to be created within the framework of the Treaty on the European Economic Community, the question will have to be examined as to what extent the above-mentioned restrictions are still permissible. Transport services of Swiss shipping companies on behalf of Swiss consignees do, however, not fall within the field of competence of the Treaty on the European Economic Community.

3. The question of the rights of cabotage

The right of cabotage is subject to widely differing regulations in the various countries.

In the Netherlands foreign shipping companies may receive unrestricted permission to participate in cabotage. They are, however, obliged to abide by the allocation of freight by the freight exchanges and by the prescribed marginal tariffs. Cabotage on the Rhine is completely free of any restrictions because this is, in the opinion of the Dutch authorities, in conformity with the spirit of the Mannheim Act. In Germany foreign ships may only in exceptional cases participate in cabotage on rivers and canals and even then they require a special permit.

They may, however, engage in cabotage in the internal German Rhine traffic and in the traffic between German Rhine ports and the canal ports as far as Hamm—Dortmund and vice-versa provided they adhere to the fixed freight rates. The Federal Government does not recognize any right of unrestricted participation in navigation implied by the Mannheim Act.

So far as Belgium is concerned foreign shipping companies may only participate in Belgian navigation on condition that they register, in common with Belgian undertakings, in a list of turns and apply the fixed freight rates.

In France foreign ships may on completion of an international journey carry out an internal one and also a second one provided there are no French ships available. The allocation of freight is based on the system of turns and the freight rates which are fixed must be adhered to.

So far as Italy is concerned, this problem does not arise.

—o—

In this instance, too, the question arises whether the restriction of the rights of cabotage is compatible with the Treaty on the European Economic Community. We rather believe that it will be necessary to arrive at a solution applicable until the end of the transitory period according to which not only traffic across frontiers but also cabotage should be free for the carriers of the participating States. This would mean that the existing restrictions in France and in Germany would have to be abolished whilst Belgium and the Netherlands would have to refrain from applying the system of permits and/or licences. Special provisions would, however, be necessary in order to meet an excessive supply of capacity in certain fields during a transitory period.

4. Fiscal policies

The Conventions of Barcelona prohibit any discriminatory fiscal policy. There is in actual fact, apart from a few exceptions, no evidence of it in inland navigation. Wherever charges are levied on internal waterways they are equally levied on national and foreign ships. Such charges are permissible by virtue of the Barcelona Conventions (Art. 7) provided they are intended to cover track costs.

Conversely, the differences between the fiscal policies of the States of the European Economic Community may exert a negative influence on a Common Market in inland navigation. Repercussions may thereby be caused through the price of fuel as well as through state subsidies to ship-building and tax reductions of various types.

The differences which exist at the present time between the regulations governing tax reductions in ship-building and in shipping in the various countries are liable to contribute to a distortion of competition. An equalization of these provisions will consequently become necessary. This, however, is rather part of the general problems of harmonization in view of the fact that as a rule not only inland navigation benefits from similar tax facilities.

5. International cartels

Traffic on the Rhine is at the present time governed by conventions and pools. Whereas a convention only fixes freight rates in order to prevent any competition based on charges a pool also provides for the sharing of all parties to the agreement in the volume of traffic. The fixed freights provided by a convention as well as the quota arrangements by means of the pools are incompatible with the division of functions within the framework of the Common Market. If a shipping company endeavours to arrive at an economically favourable level of costs by means of modernization and rational working methods the only result, at least from a short-term point of view, can be higher profits. Any reduction of freight rates which would pass on at least part of the reduction of costs to the customer is impeded by the fixed freight rates. It is true that this difficulty could be overcome by means of maximum and minimum rates. The proper functioning of such an arrangement within the framework of pools is, however, not to be expected because the fixed quota system eliminates all incentive for a reduction of prices.

The conventions and pools in Rhine navigation are a consequence of the difficult economic position in the 'thirties and the immediate post-war period. During these periods it was proved that a completely unlimited competition was bound to entail the economic ruin of many undertakings.

As a first measure intended to forestall a similar development the Freight Convention of Duisburg was concluded in 1951. By virtue of this convention parity exists between the freight rates for miscellaneous goods delivered to Germany and the sea-port rates of the German Federal Railways. As a result of an Economic Conference for Rhine Navigation at Strasbourg in 1952 the Rheinfeld pool was created in the same year for the purpose of regulating traffic to Basle. It was replaced in 1955 by the so-called Swiss Convention on Rhine Navigation. Despite this name it is, however, in actual fact a pool which covers all important goods eligible for conveyance. Similarly, the traffic of corn and coal to Strasbourg was regulated in 1955 by means of the Convention on the French Rhine Traffic. This convention, too, approaches a pool arrangement.

The most important cartel agreement is without any doubt the Convention on the International Up-Stream Coal Traffic to Germany, the so-called Kettwiger Pool of the year 1956. In conjunction with this pool a complementary agreement was concluded concerning the internal German coal traffic which is also based on a quota system. Attempts to create a pool regulating the corn traffic have, however, not been successful.

In order to safeguard the functioning of the pools certain administrative and organizational arrangements are necessary. They consist, for instance, in the case of the Kettwiger Pool, of a Plenary Assembly, a Freight Commission, and a Secretariat.

Within the framework of the pool each partner is allocated a percentage of the traffic, the so-called quota. He may also carry more than his quota; in that case, however, a quantitative adjustment among the partners becomes necessary. This adjustment may also be replaced by financial compensation.

The result of a pool agreement is not only the elimination of competition among the partners, but also their protection against competition from outsiders. For this purpose no partner to the agreement is allowed to support outsiders in any way either by providing towage or shipping space. Heavy fines may be imposed in the case of contraventions of this stipulation. Certain pool shipping companies have concluded agreements with organizations of individual owners on the basis of recommendations of the Strasbourg Conference whereby individual owners may also be included in the quota arrangements.

Despite their exclusiveness an elimination of these agreements would in our opinion not appear desirable at the present time. It would of necessity mean a return to a ruinous competition with repercussions, and by no means negligible ones, on the living conditions of the workers.

Whether the existing situation will be compatible with the provisions of the Treaty on the European Economic Community in so far as they concern the rules of competition is, however, open to considerable doubt. In contrast to many other stipulations of the Treaty it is precisely these provisions which are very clearly defined.

According to Article 85, Paragraph 1, all agreements between enterprises, decisions by associations of enterprises and any concerted practices which are likely to affect trade between Member-States and which have as their object or result the prevention, restriction or distortion of competition within the Common Market are deemed to be incompatible with the Common Market and prohibited. Although Paragraph 3 subsequently provides for certain exceptions, there is, nevertheless, reason to believe that the agreements discussed above could be declared inadmissible. The question consequently arises how the advantages inherent in these agreements could be maintained without conflicting with the spirit of the Treaty on the European Economic Community.

We could envisage a solution whereby from a long-term point of view the allocations would not be rigid but would be adapted to the structure of demand at certain intervals. A similar solution would admittedly be hampered by the shipping companies of the industrial combines which rely on strong ties with certain consignors. Within the Common Market, however, any preference expressed by a consignor for a certain enterprise should only be based on the price and the quality of the services which are offered.

In addition, certain forms of a suitable control of cartels appear to be feasible. They could be of international or supranational character. Elements of both forms may be found within the Treaty on the European Economic Community. In our opinion, however, an amendment of the Mannheim Act would be an indispensable prerequisite of such a solution.

One might also ask whether it would not be preferable to extend conventions and pools to include bodies under public law. These would then have to be subject to a regulation adapted to the field of competence of the Treaty. In connection with this it might be possible in certain circumstances to prescribe membership of shipping companies and individual owners. Organizations the membership of which would be declared obligatory for all individual owners are already being created in certain countries and they could constitute the first step on the way to such bodies under public law. It is, however, obvious that this step would by far exceed the actual conception of a control of cartels and would already imply a change of character of the cartels as such.

6. Questions of structure and organization

In Rhine navigation the functions of consignor and carrier are in many instances concentrated in one and the same hand. The majority of shares of many Rhine navigation companies is held by mines, metallurgical plants, electricity and gas works, industrial enterprises and shipping lines. This fact is bound to cause far-reaching repercussions on the functioning of a common transport market. Industrial combines with their own shipping company obviously use their ships in the first instance for the conveyance of their own goods. Consequently a considerable part of the volume of available traffic is excluded from the general competition. In order to be able at the same time to participate in the general transport market the shipping lines of the combines frequently pass on part of their reserved traffic to commercial shipping companies or individual owners. The shipping activity of the industrial combines consequently amounts, as already explained, to a kind of false transport on own account.

It is obvious that the commercial shipping companies and the individual owners work under more unfavourable conditions than those of the industrial combines. Whereas the former always have to watch their remunerativeness, the latter may, if need be, rely on an internal balance of costs within the mother enterprise. A further consequence is the fact that the shipping companies of the combines do not primarily represent the interests of inland navigation on the transport market and at discussions on conventions and pools but rather the interests of the industrial trusts. Nobody will deny that, for instance, the interests of coal mining may tend towards a different tariff policy in comparison with the interests of inland navigation.

On the other hand, inland navigation is hampered by the almost innumerable supply of transport capacity by individual owners, with an inherent tendency towards a nefarious competition, as long as the individual owners are not organized in representative federations. This fact was incidentally clearly recognized by the Strasbourg Conference in 1952, and seems to have largely contributed to the continued non-participation of the individual owners from the pool agreements.

In order to arrive at a planned competition it is consequently, in our opinion, indispensable to try to find solutions which safeguard both the independence of the existing shipping lines of the industrial combines and at the same time prevent their intervention in competition in a preferential position and on the other hand help to concentrate the widely diffused supply of transport services of the individual owners in organized units. So far as the latter are concerned, we believe, however, in contrast with the opinion expressed by the Strasbourg Conference, that they ought to be authorized to compete for freight. There is no reason why only shipping companies ought to enjoy this right. The organizations of individual owners should also be consulted on an equal footing in all discussions affecting the general shipping policy.

A prerequisite of the implementation of the above-mentioned measures in international shipping on the Rhine would be an amendment of the Mannheim Act. In this connection, certain difficulties may arise in view of the fact that not only the agreement of the states of the

European Economic Community would be required but also that of Switzerland.

7. Covering of expenditure for waterways and inland ports

Whereas on many artificial waterways and regulated rivers charges are levied which more or less approach a full covering of costs, there are other instances in which no contributions towards track costs are required. No charges whatever are, for instance, levied on the state-owned waterways of the Netherlands and, by virtue of the Mannheim Act, also on the Rhine.

The question of covering of expenditure also arises in connection with inland ports. In this case, however, a solution becomes more difficult in view of the fact that these ports are trading centres which also serve the interests of other branches of the transport industry and enterprises. In the traffic from rivers towards the open sea and particularly in the traffic from the Rhine towards the sea, inland ports also immediately serve shipping on the high seas.

In this connection, we should like to refer to our Chapter II on the distortions of competition in the first part of this report. The problem of charging track costs is, at any rate, of the greatest importance not only at national but also at international level.

8. The harmonization of freight rates

Whereas freight obligations are based on public law in the internal traffic of the states of the European Economic Community, the determination of freight rates in traffic across frontiers is not subject to any restrictions. In times of crisis these structural differences in international traffic may lead to considerably lower freight rates than on the various national routes. It goes without saying that these disparities may exert an influence on the social conditions of the personnel in inland navigation.

The problem of disparities has been very thoroughly discussed within the European Community for Coal and Steel. So far as shipping on the Rhine is concerned an agreement was signed in July 1957 by the governments of the Community. In this agreement the various states undertake to adapt their national freight rates to the level of the "representative" international freight rates. They are mainly based on existing long-term contracts between consignors and carriers, consequently in actual practice, very largely the pool rates. The adaptation of freight rates has to be carried out in consultation with the High Authority of the European Community for Coal and Steel.

The agreement has still to be ratified by the different States. For the time being it has provoked a protest by Switzerland on the grounds that this arrangement implied an infringement of the Mannheim Act.

So far as shipping west of the Rhine is concerned the European

Conference of Ministers of Transport (C.E.M.T.) has been endeavouring for years to arrive at a similar agreement. An appropriate draft was submitted to the Conference in October 1957. It provides for the compulsory creation of freight exchanges by all States. An international central committee is to be responsible for their functioning and also has to adopt all necessary measures in order to provide for the implementation of the agreement. The Government of the Netherlands has already refused to accept this agreement because it would, in its opinion, stand in the way of a genuine competition.

In view of the fact that the agreement within the European Community for Coal and Steel if it were to be ratified by the riparian states on the successful conclusion of the negotiations with Switzerland, is restricted to products of the coal and steel industries and since the draft of the C.E.M.T. concerning waterways west of the Rhine is still completely in abeyance, the problem of disparities will arise anew within the states of the European Economic Community. In this connection it is widely believed that it will not be possible simply to add national freight rates on any given route in order to arrive at the total rate in traffic across frontiers, because in international inland navigation it is in each case a question of one single transport service performed by one carrier under one given flag. The costs of such conveyance are not influenced by the national sections of the route but by the overall structure of costs of the country where the carrier is domiciled as well as by his own level of costs. Others believe, however, that international freight rates could only be based on the national systems if discrepancies which could cause extensive repercussions on social conditions were to be avoided.

We feel that a solution within the framework of the European Economic Community ought to imply an elastic method of determination of freight rates which would consequently be capable of adaptation to the economic position prevailing at any given time and whereby short-term national and international contracts would be arranged by freight exchanges. A supra-national body would be entrusted with the supervision of the development of freight rates and would also be authorized to check on long-term contracts. Minimum and maximum rates would have to be fixed in accordance with the economic position.

At any rate the determination of freight rates in traffic across frontiers should on no account be the subject of private arrangements as has been hitherto the case. Experience has shown that similar agreements become ineffectual in times of crisis. Attempts should therefore be made to arrive at an international regulation, if possible, before the culminating point of the economic boom is passed.

CONCLUSIONS

In order to forestall any doubts which may arise we wish to state quite definitely that we do not conclude this report because we have exhausted the subject. If it were only a question of dealing with problems of transport policy our report would have to be continued in instalments, practically *ad infinitum*. That, however, was never the intention and could never be. Even so, all those who participated in our work could hardly compare their experience with a pleasant journey, much as we who are part of modern transport would have liked it. It was rather an exploration of the jungle where all participants either resolutely tackled the obstacles they met and pushed them aside or, in some instances, could also do no more than circumvent them.

The experts have completed this task on behalf of the men on the job. They now return to the side-lines and leave it to those who work in transport to try to find, on the basis of this report, appropriate solutions of the problems which they encounter. Wherever gaps have been left open or wherever clear-cut definitions are lacking there exist moreover extensive possibilities of improvement of our report.

The manifold problems of the pipelines—the branch of the transport industry of the future—a statement of our policy regarding tasks and organization of the forwarding trade and a number of detail questions have, to all intents and purposes, remained untouched. Concrete statements of opinion have furthermore been omitted wherever the competent trade union organizations still have to find a common denominator for their policies. It is precisely in this field where difficulties will arise in many respects in the very near future when the framework of a common policy for newly-created extensive economic areas will have to be established, not only within the European Economic Community but also within the territory of a Free Trade Zone. We could hardly imagine that particularly in the field of transport which, by its very nature, transcends national frontiers solutions inspired by a narrow-minded provincial mentality could benefit the nations.

Although the I.T.F.—a world-wide organization—has prepared this report with reference to European conditions it does not intend to imply thereby that it is not concerned with the same problems in other parts of the world. On the contrary—the various questions would never have been considered as comprehensively as they were if we had not realized that the same problems can and will arise in only slightly different form in other regions as well.

In conclusion, we wish to express our hope that the free trade unions of transport workers may succeed in making their influence felt in the same decisive manner, whenever solutions of the many difficult economic problems of the transport industry have to be found, as they have always been able to do and will be able to do in the social field.

They will always be ready and willing to cooperate at national and international level. Occasionally, however, one cannot help wondering whether the powers that be really consider the active cooperation of the free trade unions on a basis of equality desirable. It will be very likely in this connection that the first problem of coordination will have to be solved.

Resolution (No. 37) Concerning Labour Problems Arising out of the Coordination of Transport[1]

The Inland Transport Committee of the International Labour Organisation,

Having been convened by the Governing Body of the International Labour Office,

Having met at Nervi, Genoa, in its Fourth Session from 4 to 15 December 1951,

Having noted that governments, in an attempt to ensure the best use of national resources, are promoting policies designed to achieve an effective coordination of transport and to establish conditions in which the different branches of transport can tribute efficiently and economically to the needs of the community[2]*,*

Having noted that the United Nations is considering, through its regional commission, the measures needed to promote such coordination, including the question of whether undertakings engaging in international transport operations should be subject to a permit, licence or concession to operate,

Having noted that labour costs constitute an important element in the cost of transport,

Considering that competition between transport undertakings should not be permitted to seek to take advantage of a lowering of conditions of labour and thus undermine attempts to establish a fair basis for coordination of transport, and

Considering that it is desirable to apply in the transport field the principle of "equal pay for equal work";

Adopts this fifteenth day of December 1951 the following resolution:

I

Conditions of Employment in Relation to Coordination of Transport

1. The employers' and workers' organisations concerned and the governments—in so far as the latter determine; or influence the determination of, wages and other conditions of work and employment—

[1] *Adopted on 15 December 1951 by 97 votes to 4, with 3 abstentions. For an account of the discussion leading to the adoption of this resolution, see report of the Sub-Committee on Labour Problems Arising Out of the Coordination of Transport, pp. 99-104. The original text of this resolution, as submitted by the Sub-Committee, included in paragraph 8 (b) the words "where that is impossible, eventual discharge". The replacement of this sub-paragraph by the words "material and occupational assistance to workers whose discharge is unavoidable" was proposed by the Chairman, and was adopted by the Committee by 96 votes to 3, with 2 abstentions.*

[2] *Cf. International Labour Organisation, Inland Transport Committee, Fourth Session (Genoa, 1951), Report II:* Coordination of Transport: Labour Problems *(Geneva, I.L.O., 1951).*

should make every effort to promote a greater equivalence in the conadditions of work and employment of workers engaged in the various branches of transport. This policy should aim at eliminating, or at least at mitigating progressively, the differences which exist or may exist between various branches of transport or between transport undertakings in respect of wages, social charges and the conditions of employment relating to work involving similar degrees of skill and responsibility. The best conditions of employment should be used as a guide, in so far as the particular circumstances in each country or in each branch of transport permit.

2. *Regulations concerning the operation of transport undertakings in each country should ensure the observance of fair labour standards.*

3. *For this purpose steps should be taken to ensure to the workers engaged in transport for hire or reward, wages (including allowances), hours of work and other conditions of labour which are not less favourable than those established for work of the same character in the branch of transport concerned in the district where the work is carried on—*

- *(a) by collective agreement or other recognised machinery or negotiation between the most representative organisations of employers and workers in the branch of transport concerned; or*
- *(b) by arbitration award; or*
- *(c) by national laws or regulations.*

4. *Where the conditions of labour referred to in the preceding paragraph are not regulated in the manner referred to therein in the district where the work is carried on, steps should be taken to ensure to the workers concerned wages (including allowances), hours of work and other conditions of labour which are not less favourable than—*

- *(a) those established by collective agreements or other recognised machinery of negotiation, by arbitration, or by national laws or regulations, for work of the same character in the branch of transport concerned in the nearest appropriate districts; or*
- *(b) the general level observed in the branch of transport concerned by employers whose general circumstances are similar.*

5. *In cases in which operators are subject to the grant of a permit, licence or concession to undertake the transport of passengers or of goods, the observance of the provisions relating to wages (including allowances), hours of work and other conditions of labour specified in paragraphs 3 and 4 above should be a condition of the grant or retention of the permit, licence or concession, where other regulations do not already exist for ensuring the observance of these standards.*

Transport on Own Account

6. The competent authority in each country, after consulting the employers' and workers' organisations concerned, should consider the possibility of applying the provisions of paragraphs 3, 4 and 5 above to those persons whose major occupation is in transport and who are employed by undertakings carrying out transport on own account, bearing in mind the special regulations or collective agreements applicable to the various industries to which these workers may belong.

Social Consequences of Coordination

7. No measures for co-ordination should be adopted without taking into account their social consequences.

8. In cases in which measures aiming at the coordination of transport are liable adversely to affect the workers concerned, measures should be taken either by the competent authority, after consultation with the employers' and workers' organisations concerned, or by agreement between these organisations, regarding either—

(a) transfer within the same occupation or, where necessary, to another occupation, in particular by assisting those workers who are obliged to change their occupation or place of work and by providing vocational training for workers compelled to change their occupation; or

(b) material and occupational assistance to workers whose discharge is unavoidable.

During the negotiations which will lead to adoption of one of the measures referred to above, attention should be specially drawn to the advantage there would be in considering whether special measures concerning the maintenance of certain acquired rights could not be adopted.

Cooperation of Employers' and Workers' Organizations Concerned

9. The employers' and workers' organisations concerned should be closely associated an an equitable basis with the work of bodies dealing with the coordination of transport either through participation in them or by means of consultation.

Supervision

10. The appropriate authorities or the contracting parties should organise labour inspection in transport undertakings in an efficient manner with a view to ensuring observance of the legal or other regulations concerning conditions of work and the protection of transport workers.

11. For the purpose of ensuring proper control of the enforcement of fair labour standards, the appropriate authorities or the contracting parties should, whenever necessary, cause to be kept, preserved and placed at their disposal records concerning, in particular, wages, allowances, social charges, hours of work, weekly rest, rest on public holidays and overtime. These records should not, however, involve undue formalities or administrative costs

Sanctions

12. Adequate sanctions should be applied for failure to observe the provisions relating to wages (including allowances), hours of work and other conditions of labour. Such sanctions should include, where appropriate, the cancellation of the permit, licence or concession.

II

13. The Governing Body of the International Labour Office is invited—

(a) to authorise the Director-General to communicate to the United Nations the report of the Sub-Committee on Labour Problems Arising Out of the Coordination of Transport and the conclusions of the Committee on the subject; and

(b) to instruct the Director-General to continue to follow the discussions in the United Nations and in other international organisations relating to the coordination of transport with a view, whenever necessary, to bringing the social aspects of the question to the attention of those concerned.